DIFFERENT

Other books by Frans de Waal

Mama's Last Hug (2019)

Are We Smart Enough to Know How Smart Animals Are? (2016)

The Bonobo and the Atheist (2013)

The Age of Empathy (2009)

Primates and Philosophers (2006)

Our Inner Ape (2005)

My Family Album (2003)

The Ape and the Sushi Master (2001)

Bonobo: The Forgotten Ape (1997)

Good Natured (1996)

Peacemaking Among Primates (1989)

Chimpanzee Politics (1982)

DIFFERENT

GENDER THROUGH THE EYES OF A PRIMATOLOGIST

Frans de Waal

With drawings & photographs by the author

W. W. NORTON & COMPANY
Independent Publishers Since 1923

For information about permission to reproduce selections from this book, write to
Permissions, W. W. Norton & Company, Inc., 500 Fifth Avenue, New York, NY 10110

For information about special discounts for bulk purchases, please contact
W. W. Norton Special Sales at specialsales@wwnorton.com or 800-233-4830

All drawings by the author. All photographs by the author, unless otherwise indicated.

Manufacturing by Lake Book Manufacturing
Production manager: Devon Zahn

Library of Congress Cataloging-in-Publication Data
Names: Waal, F. B. M. de (Frans B. M.), 1948– author, illustrator.
Title: Different : gender through the eyes of a primatologist / Frans de Waal with
    drawings & photographs by the author.
Description: First edition. | New York, NY : W. W. Norton & Company, [2022] |
    Includes bibliographical references and index.
Identifiers: LCCN 2021045201 | ISBN 9781324007104 (hardcover) |
    ISBN 9781324007111 (epub)
Subjects: LCSH: Sex differences. | Sexual dimorphism (Animals) | Sexual behavior
    in animals.
Classification: LCC QP81.5 .W33 2022 | DDC 612.6—dc23
LC record available at https://lccn.loc.gov/2021045201

W. W. Norton & Company, Inc., 500 Fifth Avenue, New York, N.Y. 10110
www.wwnorton.com

W. W. Norton & Company Ltd., 15 Carlisle Street, London W1D 3BS

1 2 3 4 5 6 7 8 9 0

For Catherine, who makes all the difference

CONTENTS

DIFFERENT

INTRODUCTION

The saddest day of my career began with a phone call telling me that my favorite male chimpanzee had been butchered by two rivals. Having hurried on my bike to Royal Burgers' Zoo, in the Netherlands, I found Luit sitting in a puddle of blood, leaning his head dejectedly against the bars of his night cage. Normally aloof, he heaved the deepest sigh when I stroked his head. It was too late, though. He died the same day on the operating table.

Rivalry among male chimps can grow so intense that they kill each other, and not only at the zoo. There are now a dozen reports of high-ranking males slain in the wild during the same sort of power struggles. While jockeying for the top spot, males opportunistically make and break alliances, betray each other, and plot attacks. Yes, plot, because it was no accident that the assault on Luit took place in the night quarters where three adult males were kept apart from the rest of the colony. Things might have unfolded differently out on the large forested island of the world's best-known chimpanzee colony. Female chimps don't hesitate to interrupt clashes among male contenders. While Mama, the alpha female, couldn't keep the males from politicking, she did draw the line at bloodshed. Had she been present on the scene, she'd no doubt have rallied her allies to step in.

Luit's untimely death affected me deeply. He had been such a friendly character, who as leader had brought peace and harmony. But on top of that, I was deeply disappointed. Until then, the battles I had witnessed had always ended in reconciliation. Rivals would kiss and embrace after each skirmish and were perfectly capable of handling their disagreements. Or so I thought. Adult male chimps act like friends most of the time, grooming each other, and roughhousing in fun. The disastrous fight taught me that things can also spiral out of control and that the same males are capable of intentionally killing each other. Fieldworkers have described assaults in the forest in similar tones. They seem deliberate enough to speak of "murder."

The high-intensity aggression of male chimps has a female equivalent. The circumstances that trigger female anger are quite different, though. Even the biggest male knows that every mother will turn into a raging hurricane if he lifts a finger against her progeny. She will become so undaunted and fierce that nothing will stop her. The ferocity with which a mother ape defends her young exceeds that with which she defends herself. Maternal protectiveness is such a universal mammalian trait that we joke about it, such as when U.S. vice-presidential candidate Sarah Palin called herself a Mama Grizzly. Mindful of this reputation, Gary Larson drew a cartoon in which a businessman carrying a briefcase enters an elevator with a large and a small bear standing in the back. The caption reads: "Tragedy struck when Conroy, his mind preoccupied with work, stepped into the elevator—directly between a female grizzly and her cub."

The greatest fear of *fandis* in the jungles of Thailand—hunters who in the old days captured wild elephants for timber labor—was not that they'd snare a tusker. A large bull in the ropes posed less acute danger than a small calf captured while its mother was within hearing range. Quite a few *fandis* have lost their lives to enraged elephant cows.[1]

In our species, a mother's defense of her children is so predictable that, according to the Hebrew Bible, King Solomon counted on it. Faced with two women who both claimed to be the mother of a baby, the king asked for a sword. He proposed to split the baby so that each woman could have half of it. While one woman accepted the verdict, the other protested and pleaded that the baby be given to the other. This is how the king knew who the real mother was. As the British detective writer Agatha Christie put it, "A mother's love for her child is like nothing else in the world. It knows no law, no pity, it dares all things and crushes down remorselessly all that stands in its path."[2]

While we admire mothers who take their children's side, we hold a dim view of human male combativeness. Boys and men often instigate confrontations, act tough, hide vulnerabilities, and seek danger. Not everyone likes men for this, and some experts disapprove. When they say that "traditional masculinity ideology" fuels men's behavior, they hardly mean it as a compliment. In a 2018 document, the American Psychological Association defined this ideology as revolving around "anti-femininity, achievement, eschewal of the appearance of weakness, and adventure, risk, and violence." The APA's attempt to save men from this ideology revived debate about "toxic masculinity" but also triggered backlash over its blanket denunciation of typical male behavior.[3]

It's easy to see why male and female patterns of aggression are valued so differently: only the first creates trouble in society. Horrified by the death of Luit, I don't want to depict male rivalry as an innocuous pastime. But who says it's a product of ideology? A huge assumption is being made here, which is that we are the masters and designers of our own behavior. If this were true, shouldn't it stand apart from that of other species? But it hardly does. In most mammals, males strive for status or territory whereas females vigorously defend their young. Whether we approve or disapprove of such behavior, it's not hard to see how it evolved. For both sexes, it has always been the ticket to a genetic legacy.

Ideology has little to do with it.

Sᴇx ᴅɪꜰꜰᴇʀᴇɴᴄᴇꜱ ɪɴ animal and human behavior raise questions that lie at the heart of almost any debate about human gender. Does the behavior of men and women differ naturally or artificially? How different are they really? And are there only two genders, or are there more?

But before I dive into this topic, let me make clear why I am interested in it and where I stand. I am not here to justify existing human gender relations by describing our primate heritage; nor do I think that everything is fine as it is. I recognize that the genders are not now and have never been equal for as long as we can remember. Women get the short end of the stick in our society and in almost every other one. They have had to fight for every improvement, from the right to education to voting rights, and from legalized abortion to equal pay. These aren't little improvements. Some rights have been secured only recently, some are still being resisted, and some were achieved but have come under fresh attack. I see all this as highly unfair and consider myself a feminist.

Disdain for the innate abilities of women has a long tradition in the West going back at least two millennia. It's the way gender inequality has always been justified. Thus the nineteenth-century German philosopher Arthur Schopenhauer thought that all their lives women remain children, who live in the present, whereas men have the ability to think ahead. Another German philosopher, Georg Wilhelm Friedrich Hegel, thought that "men correspond to animals, while women correspond to plants."[4] Don't ask me what Hegel meant, but as noted by the British moral philosopher Mary Midgley, when it comes to women, the heavyweights of Western thought have produced extraordinarily silly reflections. Their usual divergence of opinion is nowhere to be found: "There cannot be many matters on which Freud, Nietzsche, Rousseau and Schopenhauer agree cordially both with each other and with Aristotle, St. Paul and St. Thomas Aquinas, but their views on women are extremely close."[5]

Even my beloved Charles Darwin didn't escape the trend. In a letter to Caroline Kennard, an American women's rights advocate, Darwin opined about women, "There seems to me to be a great difficulty from the laws of inheritance in their becoming the intellectual equals of man."[6]

All this in an era when disparities in education could easily account for the proposed intellectual contrasts. As for Darwin's "laws of inheritance," all I can say is that I've devoted my entire career to the study of animal intelligence and never noticed a difference between the sexes. We have brilliant individuals and not-so-brilliant ones on both sides, but hundreds of studies by myself and others have revealed no cognitive gaps. While there is no shortage of behavioral contrasts between primate males and females, their mental capacities must have evolved in tandem. In our species, too, even the cognitive domains traditionally associated with one gender and not the other, such as mathematical ability, prove indistinguishable by gender if tested on a large enough sample.[7] The whole idea of one gender being mentally superior receives no backing from modern science.

A second issue that needs to be cleared up is the stereotypical view of our fellow primates that is sometimes used to defend inequalities in human society. In the popular imagination a male monkey boss "owns" the females, who spend their lives making babies and following his orders. The chief inspiration for this view was a baboon study of one century ago that, as I will explain, had major flaws and gave rise to a dubious metaphor.[8] Unfortunately, it hit the public like a barbed arrow that proved impossible to dislodge despite all the contrary information gathered since then. That male supremacy is the natural order was promulgated over and over by a slew of popular writers in the previous century, while a 2002 book, entitled *King of the Mountain,* by the American psychiatrist Arnold Ludwig still maintains:

Most humans have been socially, psychologically, and biologically programmed with the need for a single dominant male figure to govern their communal lives. And this programming corresponds closely to how almost all anthropoid primate societies are run.[9]

One of my goals here is to disabuse readers of this notion of the obligatory male overlord. The primate study at its origin concerned a species that we are not particularly close to. We belong to a small family of apes (large tailless primates), not of monkeys like baboons. By studying our next of kin, the great apes, a more nuanced picture emerges, one in which males exert less control than imagined.

While it is undeniable that male primates can be bullies, it's also good to realize that they didn't gain their aggressiveness and size advantage in order to dominate females. This is not what their life is about. Given the ecological demands, females evolved to be the perfect size. Their bodies are optimal given the foods they gather, the amount of traveling they do, the number of offspring they raise, and the predators they elude. Evolution has pushed males to deviate from this ideal so as to better fight each other. [10] The more intense the competition among them, the more impressive their physical features. In some species, such as the gorilla, the male is twice the female's size. Since the whole point of male fighting is to get close to females with whom to reproduce, harming them or taking away their food is never the males' goal. In fact, most female primates enjoy a great deal of autonomy, foraging all day for themselves and socializing with each other, while the males are peripheral to their existence. The typical primate society is at heart a female kinship network run by older matriarchs.

We heard the same reflection when *The Lion King* was newly released. In the movie, the male lion is depicted as the boss—because most people cannot conceive of a kingdom any other way. The mother of Simba, the cub destined to become the next king, hardly plays any role at all. However, while it is true that lions are bigger and stronger than lionesses, they hold no central position in the pride. The pride is essentially a sisterhood, which does the bulk of hunting and offspring care. Male lions stay for a couple of years before they are kicked out by incoming rivals. As Craig Packer, one of the world's leading lion experts, puts it, "Females are the core. The heart and soul of the pride. The males come and go." [11]

While comparing ourselves with other species, the popular media feature a surface reality. The deeper reality, however, can be quite different. It may reflect substantial sex differences but not necessarily the ones we expect. Moreover, many primates have what I call *potentials*, which are capacities that are rarely expressed or hard to see. A good example is female leadership, such as I described in my last book, *Mama's Last Hug,* for the longtime alpha female at Burgers' Zoo. Mama was absolutely central to social life, even though, measured by the outcome of fights, she ranked below the top males. The oldest male, too, ranked below them but was equally central. Understanding how these two aging apes together ran a large chimpanzee colony requires looking beyond physical dominance and recognizing who makes the critical social decisions. We need to distinguish political power from dominance. In our societies, no one confuses power with muscularity, and the same holds true for those of other primates.[12]

Another potential is the caretaking capacity of male primates. We sometimes get a glimpse of it after a mother's death, when all of a sudden an orphan whimpers for attention. Adult male chimpanzees in the wild have been known to adopt a little one and lovingly care for it, sometimes for years. The male will slow down his travel for the adopted youngster, search for him if he's lost, and be as protective as any mother. Since scientists tend to stress typical behavior, we don't always dwell on these potentials. Still, they bear on human gender roles given that we live in a changing society, which tests the limits of what our species is capable of. There is every reason, therefore, to see what we can learn about ourselves from comparisons with other primates.[13]

Even those who doubt evolutionary explanations, and think that the same rules don't apply to us, will have to admit to one basic truth about natural selection. No person currently walking the earth could have gotten here if it weren't for ancestors who survived and reproduced. All our ancestors conceived children and raised them successfully or helped others raise theirs. There are no exceptions to this rule because those who failed to do so are ancestors to no one.

Their genes are absent from the gene pool.

Mᴏᴅᴇʀɴ ꜱᴏᴄɪᴇᴛʏ ɪꜱ ready for a correction of gender differences in power and privilege. Women cannot accomplish this alone, though. Gender roles are so intertwined that both men and women will need to change at the same time. Some of these adjustments are already under way. I see a younger generation doing things quite differently than my own, such as the greater involvement of men in parenting and the incursions of women into male-dominated jobs. The way forward is to get men on board. This is why I bristle at generalizations, such as those blaming men for all that's wrong in the world. Calling certain expressions of masculinity "toxic" is not my idea of feminism. What is the point of stigmatizing an entire gender? I agree with American actress Meryl Streep, who saw this as unnecessary: "We hurt our boys by calling something toxic masculinity. Women can be pretty fucking toxic. . . . It's toxic people."[14]

It's nearly impossible for us to know the origin of most human gender differences in everyday life. After all, our culture puts constant pressure on both men and women. Everyone is supposed to fall in line and fit the rules of masculinity and femininity. Is this how we create gender, and has gender superseded biological sex? This cannot be the whole answer, though. Other primates aren't subject to our gender norms yet often act like us, and we like them. While their behavior, too, may follow social norms, these norms would derive from *their* culture, not from ours. More likely, similarities between their behavior and ours hint at a shared biology.

Other primates hold up a mirror to ourselves, which allows us to see gender in a different light. They aren't us, however, and so they offer a comparison, not a model for us to emulate. I add this qualification because factual statements are sometimes taken as normative, which they aren't. People find it hard to consider descriptions of other primates without relating them to themselves. They praise them for acting in ways they approve of and get upset if they do things they hate to see. Since I study two kinds of ape with radically different relations between the sexes, I immediately pick up on these reactions in my audiences when I speak about them. People sometimes react as if my descriptions were endorsements. Whenever I discuss chimpanzees, they think I must be a fan of male power and brutality. As if I thought that it would be great if men acted like this! And when I present the social life of bonobos, my listeners are sure that I delight in eroticism and female control. In reality, I like both bonobos and chimpanzees and find them equally fascinating. They reveal different sides of ourselves. We have a bit of each ape inside us, while in addition we've had several million years to evolve our own unique traits.

As an example of people getting upset, let me take you back to a day when, as a young man, I lectured at Burgers' Zoo about the chimpanzees that I was studying there. I spoke to a great variety of audiences, from the bakers' guild to the police academy to schoolteachers and children. They all liked my stories, until the day I addressed a group of women lawyers. They were distinctly unhappy with my message and even called me "sexist," a charge that had only just become common currency. But how could they conclude this given that I hadn't said a word about human behavior?

I had described how male and female chimps differ. Males perform spectacular bluff displays that express their drive for power. They are strategical, always planning their next move. The females, on the other hand, spend most of their time grooming and socializing. They focus on relationships and family. I was also proud to show pictures of our latest baby boom in the colony. But my listeners were not in the mood to coo over baby apes.

Afterward the lawyers asked me how I could be so sure that males dominate females. Why couldn't it be the other way around? they asked. Perhaps I had the wrong idea of dominance, they suggested. Although I said I had seen the males win fights, in reality, they said, it might very well have been the females who won. Having spent every day and thousands of hours with the chimps, I was being corrected by people who could barely tell a chimp from a gorilla! Even though my field of study has no shortage of female experts, I have never heard chimpanzees described as anything other than male-dominated. This refers only to physical superiority, which is a narrow take but meaningful nonetheless. Male chimpanzees are heavier than females and are shaped like bodybuilders, with massive arms and shoulders and a thick neck. They are also armed with long canine teeth, almost like a leopard's, which the females lack. The females are no match. The only exception occurs when females band together.

Later that same day, during a visit to the chimp island, the group of lawyers came around a bit when they saw with their own eyes a few incidents that confirmed my point. But it did nothing to improve their mood.

Later in life, when I worked with bonobos and lectured about them, the opposite happened. Chimpanzees and bonobos are both anthropoid apes, both genetically extremely close to us, but they are surprisingly different in behavior. Chimpanzee society is aggressive, territorial, and run by males. Bonobos are peaceful, sex-loving, and female-dominated. How much more unalike can two apes get? Bonobos give the lie to the idea that knowing more about our fellow primates is bound to reinforce gender stereotypes. As the scientist who dubbed them the "make love not war" primates, I opened my first popular article on this species with the line: "At a juncture in history during which women are seeking equality with men, science arrives with a belated gift to the feminist movement." This was back in 1995.[15]

Audiences applaud bonobos. They love them, feeling they bring light in a time when biology strikes them as dark. The novelist Alice Walker dedicated *By the Light of My Father's Smile* to our close kinship with bonobos, and the *New York Times* columnist Maureen Dowd once mixed political commentary with praise of bonobos' egalitarian ethos. Bonobos have been called the "politically correct primate" both for the dominance reversal between males and females and for their incredibly varied sex lives. They do it in all partner combinations, not only male and female. I'm always glad to talk about our hippie relatives, but I don't think evolutionary comparisons should be biased by wishful thinking. We cannot just go around the animal kingdom and pick and choose which species we like the best.

If we have two ape relatives that are equally close to us, then they are equally relevant to our discussions about the relation between the sexes. This book will emphasize both of them, even though chimpanzees have been known to science for much longer and are better studied. I will pay less attention to other primates, such as monkeys, that are more distant from us.

THE SUBJECT OF gender differences arouses emotions one way or the other. It is an area where everyone holds strong opinions, which is not something we're used to with regard to animals. Primatologists try not to judge. We don't always succeed, but we never classify behavior as right or wrong. Our work includes interpretation, which is unavoidable, but you won't hear us drop a term like *obnoxious* for male behavior or call the females of a species *mean*. We take behavior as it comes. This attitude has a long tradition among naturalists. Even though the male praying mantis literally loses his head during copulation, no one is going to blame the female. In the same way and for the same reason, we don't judge the male hornbill, who brings lumps of clay to allow his mate to seal herself inside the nest cavity for weeks. All we do is wonder why nature works this way.

This is also how a primatologist looks at society. We don't worry about the desirability of behavior but rather try to describe it as best we can. It's a bit like that video spoof in which David Attenborough, the British naturalist and TV personality, narrates our species' mating rituals. Over footage of frat boys swilling beer in a Canadian bar, Attenborough's soothing voice intones that "the air is heavy with the scent of females" and "each male tries to show how strong and dexterous he is." The video ends with a "winner" in bed with one of the women, where she takes charge of the situation.[16]

Is it sexist? Only if you believe that any allusion to sex-typical behavior implies a political stance. We live in a time when some people systematically hype sex differences as if they were everywhere, while others try to erase them by depicting them as meaningless. The first group will alight on minor differences in spatial memory, moral reasoning, or anything else and blow them out of proportion. Their conclusions are often amplified by the media, which will turn a variation of a few percentage points in favor of either sex into a black and white difference. Some writers will even say that men and women hail from different planets. Another group will do the opposite: they will soften any statement about how men and women differ. "It doesn't apply to all of us," they'll claim, or "It's a product of the environment." Their keyword is *socialization*, as in "men are socialized to compete" or "women are socialized to care for others." They purport to know where differences in behavior come from and that it's definitely not biology.

One of the early advocates of the latter position is the American philosopher Judith Butler, who considers "male" and "female" to be mere constructs. In a seminal 1988 article, she stated, "Because gender is not a fact, the various acts of gender create the idea of gender, and without those acts, there would be no gender at all."[17] Hers is an extreme position with which I can't agree. Nevertheless, I do consider *gender* a useful concept. Every culture has different norms, habits, and roles for the sexes. Gender refers to the learned overlays that turn a biological female into a woman and a biological male into a man. It's true that we are thoroughly cultural beings. I'd even go further and argue that the concept of gender may suit other primates as well. Apes reach adulthood by about sixteen years of age, which gives them ample time to learn from others. If this alters their sex-typical behavior, we should speak of genders in their case as well.

Gender also covers identities that don't correspond to biological sex, such as those of transgender men and women. And there are other exceptions, such as when the anatomical or chromosomal sex of a person is hard to classify or when people don't identify as either one or the other gender. Nevertheless, for the majority of people, gender and sex are congruent. Despite their different meanings, these two terms remain joined at the hip. Thus, a discussion of gender differences automatically covers sex differences, and vice versa.

Science has ignored sex differences for the longest time, but this has begun to change. One reason is that the neglect has damaged health care.[18] Women used to be diagnosed and treated like men—small men. Ever since Aristotle's observation that "the female is, as it were, a mutilated male," medicine has taken the male body as its gold standard. The only modification the female body required, it was thought, was a lower dose of whatever medication had been developed for men.[19]

Male and female bodies are far from the same, however. Some of the differences are merely structural. For example, women are more likely than men to suffer serious injuries in car accidents, which may be due to a difference in bone density or because the car industry still uses crash test dummies based on male bodies, which have a different weight distribution than female bodies.[20] The differences extend to sex-specific conditions (such as those related to uterus, breasts, and prostate) and other health vulnerabilities. In 2016 the National Institutes of Health called on medical scientists in the United States to always include both sexes in their research. *NIH Policy on Sex as a Biological Variable* covers all vertebrates, such as mice, rats, monkeys, and humans. Many diseases are sex-biased. For example, women have a higher chance than men of developing Alzheimer's, lupus, and multiple sclerosis. In contrast, men have a higher incidence of Parkinson's and autism spectrum disorder. Overall, women are the sturdier sex and live longer than men, a difference found in most mammals. These differences have little to do with Butler's "idea of gender" and everything to do with sex at birth.[21]

Primatologists have no reason to downplay sex. I must have listened to a thousand lectures at primatological conferences, yet I have never heard anyone say, "You know, I've followed male and female orangutans through the forest and find their behavior to be strikingly similar." The speaker would be laughed out of the room given how clear behavioral sex differences are in most primates. Moreover, we love these differences. They are our bread and butter. It's what makes the social life of primates so fascinating. The males have one agenda, the females another, and our task is to figure out the interplay between the two. Males and females sometimes have conflicting interests, but since neither sex can win the evolutionary race without the other, their agendas always intersect at some point.

Not that my comparisons yield easy answers. Some purported sex differences have proven impossible to confirm, while those that do exist are often less straightforward than imagined. In setting our species against a primate backdrop, I will make use of a rich literature on human behavior. I do so selectively and as a relative outsider. My main bias is that I don't trust human self-reports. Asking people about themselves has become fashionable in the social sciences, but I prefer to go back to an earlier time when we still tested and observed actual behavior, such as how children play in the schoolyard or how athletes react when they have won or lost. People's behavior is so much more informative and honest than what they say about themselves! It's also easier to compare to primate behavior.[22]

My discussion of human gender relations will overlook some important issues. Since primatological observations are my starting point, I will consider only related human behavior, thus leaving aside areas for which we have no animal parallels, such as economic disparities, household labor, access to education, and cultural rules for attire. My expertise is unable to shed light on those issues.

WHETHER THE PUSH for gender equality will succeed doesn't hinge on the outcome of the eternal debate about real or imagined sex differences. Equality doesn't require similarity. People can be different and still deserve exactly the same rights and opportunities. So an exploration of how the sexes differ in both human and other primates in no way validates the status quo. I sincerely believe that the best way to achieve greater equality will be to learn more about our biology instead of trying to sweep it under the rug. In fact, the whole reason we are having this conversation is thanks to a small biological invention that radically transformed society.

The estrogen/progestin tablet that prevents ovulation (the release of eggs from the ovaries) has been so impactful that it is simply known as "the Pill." No other pill has this privilege. Its introduction in the 1960s was a watershed moment as it permitted the uncoupling of sex from procreation. People could now have smaller families, or no families at all, without having to forgo intercourse. Effective birth control gave us the sexual revolution, from Woodstock to the gay rights movement. In one fell swoop, it called into question traditional morals regarding pre- and extramarital sex and also many other expressions of sexuality. Feminists began to see women's pursuit of sexual pleasure as part of achieving greater independence. Changes in gender roles, too, can be traced back to the introduction of the Pill. In a society in which women performed the bulk of childcare, having no children or only a few changed their need to stay home. In the 1970s, after moral restrictions on the Pill (such as its denial to unmarried people) were lifted, women began to massively enter the workforce.

I wouldn't be here to discuss the Pill if it had existed around the time of my conception. My parents didn't want a large family, but they lived in a part of the Netherlands, known as the Catholic South, where the Church held outsize sway. It opposed any kind of family planning. A familiar story in our family is how my mother, not long after she had delivered her sixth child, got angry at a priest visiting our home. Sitting comfortably with coffee and a cigar, the priest had casually brought up "the next one." He didn't get to finish his coffee and was sent packing. I got no more new siblings after that. Attitudes were already changing before the Pill, but once it arrived, it made everything easier. In the decades that followed, the size of families in our region dropped precipitously.

Thus, a bit of tinkering with human biology reshaped the playing field, which goes to show that biology doesn't have to be the enemy. I personally look at it as our friend. Humanity needed the Pill because the most logical alternative way to prevent pregnancy doesn't quite work for us. We could simply have stopped having sex or at least abstained for intermittent periods. But this is too much to ask of the lusty apes that we are. Also solutions that require men to stop and think and put on a condom before they act have proven unreliable. This is partly due to the passion of the moment and partly because it leaves things up to the least concerned gender. The Pill changed all that. Human biology required a biological answer. It still does, even if we have begun to worry about the Pill's side effects on mood and mental health.

We are animals, and within this category we belong to the primate order. Just as we share at least 96 percent of our DNA with chimpanzees and bonobos (the exact percentage is under debate), we share our socio-emotional makeup with them. How much we have in common is not sure, but far less separates us than we're led to believe. While many academic disciplines love to stress human uniqueness and put us on a pedestal, that perspective is increasingly out of touch with modern science. If humankind is a floating iceberg, they ask us to fixate on the shiny little tip of our differences with other species while overlooking the vast commonality that hides below the surface. Biology, medicine, and neuroscience, on the other hand, prefer to contemplate the whole iceberg. They know that, even if the human brain is relatively large, it barely differs from a monkey brain in terms of its structure and neural chemistry. It has the same parts and works the same way.

A funny thing happened to me once during an interview on Norwegian national television. While I was discussing the evolution of empathy, the interviewer asked me, almost as an aside, "How's Catherine doing?" This shocked me. If people ask me about the ape personalities in my books, that's okay—I always have a story to tell about them. Catherine, however, is my wife. So I said, "She's doing well," hoping we'd move on. But then the interviewer asked, "How old is she now?" I answered,

"She's about my age, why?" Surprised, the interviewer said, "Oh, they get *that* old!" That's when it hit me that she thought Catherine was one of my study subjects.

I suddenly realized the source of this misunderstanding. After all, I had dedicated my latest book, "For Catherine, my favorite primate."

1

TOYS ARE US

How Boys, Girls, and Other Primates Play

One morning, through my binoculars, I watched Amber walk out onto the island in an oddly bent posture, hobbling on one arm and two legs. She was clutching the head of a soft broom against her belly and supporting it with one hand, exactly the way a mother ape holds a newborn that is too small and weak to cling on its own. Amber—named after the color of her eyes—was an adolescent female in the chimpanzee colony of Burgers' Zoo. One of the keepers must have accidentally left the broom behind, and Amber had pulled out the handle. She occasionally groomed the brush and walked around with it positioned on her lower back like a mother carrying an older offspring. At night, she'd curl up with it in her straw nest. She kept the broom close for weeks. Instead of mothering other females' infants, she now had one of her own, except it wasn't real.

When apes are given dolls to play with, one of two things will happen. If a young male gets a hold of it, he may tear it apart—mainly out of curiosity to see what's inside, but sometimes due to competition. When two young males both pull at a doll, each may end up with part of it. In the hands of males, toys rarely enjoy a long life. If a female gets hold of a doll, on the other hand, she will soon adopt it and treat it gently. She will take care of it.

A juvenile female chimpanzee named Georgia once entered an indoor area with a teddy bear she had been carrying around for days. I knew her well and wanted to see if she'd be willing to let me hold her bear. I stretched out an open hand in begging fashion, a gesture that chimps themselves use. We had bars between us, and Georgia was conflicted. She kept the bear away from me. So I sat down on the floor to show her that I wasn't going to walk off with it. She then pushed the bear toward me while firmly holding on to a leg. She allowed me to inspect and talk to it, closely watching me. By the time I pushed the bear back to her, we had bonded over this act of trust, and she tightly cuddled her bear while staying next to me.

The primate literature is full of apes in human care—almost all females—who nurture dolls they have been given. They drag them around, carry them on their backs, and hold their mouth against a nipple as if they were nursing; or like Koko the sign-language gorilla, they kiss their dolls goodnight one by one, after which they reenact a round of all of them kissing each other.[1]

Another language-trained ape, the chimpanzee Washoe, once used her doll as a guinea pig. Upon noticing that a new doormat had been installed in her trailer, she jumped back in horror. She grabbed her doll and, from a safe distance, tossed it onto the mat. She monitored the situation intently for a few minutes, to see if anything happened to her doll, then snatched it off the mat and inspected it carefully. Concluding that it was unharmed, she calmed down and dared cross the mat.[2]

It is said that people socialize boys and girls through toy choices. By pushing our own prejudices onto them, we mold their gender roles. The idea is that children are blank slates filled in by their environment. While it's true that many aspects of gender are culturally defined, not all of them are. Since toys are central to this debate, they offer an excellent starting point for discussion. The toy industry tells us what our daugh-

When children's toys were given to monkeys, the wheeled vehicles ended up mostly in the hands of young males and the dolls in those of young females. The difference was driven by a lack of male interest in dolls.

ters and sons need, but even if we were to buy up a whole toy store, it would still be up to our children which toys they pick. This is the beauty of play: it's up to the player. It's best just to watch children entertain themselves with their reenactments and imagination and to remain open to the possibility that, instead of us shaping them, it could be the other way around.

Judith Harris, an American maverick psychologist, saw parental influence as a mere feel-good illusion. In her 1998 book *The Nurture*

Assumption: Why Children Turn Out the Way They Do, she surmised, "Yes, parents buy trucks for their sons and dolls for their daughters, but maybe they have a good reason: maybe that's what the kids want."[3]

As I watched amber with her broom-child, it was evident that she wanted a doll. Is this typical of female primates? When scientists have tested toys on monkeys, their choices have proved anything but sex-neutral. In the first such experiment, conducted twenty years ago at the University of California in Los Angeles, Gerianne Alexander and Melissa Hines gave vervet monkeys a police car, a ball, a plush doll, and a few other toys. Admittedly, this was a contrived setup, full of assumptions of what these objects might mean to monkeys. I prefer experiments inspired by the animals' actual behavior instead of by our anthropocentric tendency to throw human issues at them. But let's see what they found.

The monkeys mimicked the sex-based preferences of human children. Transportation toys, such as cars, were handled more by the males, who moved them along the ground. Males also liked the ball. Dolls, on the other hand, were carried more by the females, who'd pick them up to hold them tight or take a close peek at their genital region. The latter fits the curiosity of monkeys about the genitals of newborns. It's not unusual for females to gather around a new mother to spread her wriggling infant's legs, prodding, pulling, and sniffing between them amid a chorus of soft grunts and lip-smacks. They seem to agree on the importance of this part of the body. Primates have been doing this for ages, long before we invented "gender reveal" parties.[4]

That UCLA study did not present all toys at the same time, so the monkeys couldn't truly make a choice. All we know is how long they played with each kind of toy. A second study, using rhesus monkeys at the Field Station of the Yerkes National Primate Research Center near Atlanta, addressed this shortcoming. Since I work there, I walk by these monkeys every day. Year round, they live outdoors in large fenced-in corrals, where they engage in noisy squabbles, grooming get-togethers, and wild play sessions. Although they have lots of things to do, new toys catch their attention. Kim Wallen, a colleague of mine at Emory University, and his graduate student Janice Hassett, gave a group of 135 monkeys two kinds of toys to see which ones they'd pick. They provided these toys simultaneously: soft plush toys, such as dolls, and wheeled ones, such as cars.[5]

Male monkeys went for the wheeled toys. They were more single-minded than the females, who liked all the toys, including the cars. Due to the males' indifference to plush toys, most of those ended up in female hands. Children show a similar pattern, with boys having more pronounced toy preferences. A common explanation is that boys are uneasy about appearing feminine, whereas girls worry less about appearing masculine. But absent evidence that monkeys care about gender perception, it's unlikely that they feel the same uneasiness that boys are said to have. The reality may be more straightforward: dolls may just not appeal to most boys and male primates.

The setup of these experiments was odd because they presented monkeys with artificial items that they weren't accustomed to. This drawback applied especially to the trucks. Colorful vehicles made out of plastic or metal don't look like anything in their natural habitat. Were male monkeys fascinated by movable objects that invite action, such as balls and cars? Males have a high energy level and like physical play. That females played with huggable plush toys is easier to explain. The dolls had a body, head, and limbs, which made them superficially like babies or animals. Female monkeys will spend the rest of their lives caring for infants, whereas males won't.[6]

I never played with dolls even though my mother always kept a few around for my brothers and me. I was fond of my large stuffed bulldog but never slept with it and sometimes sent it flying while practicing my boxing skills. My typical play items were crayons and paper because I loved drawing, and building materials, such as an Erector Set and electric toy trains. My greatest interest by far, however, was animals. I don't know how or when this started, but from a very young age, I collected frogs, grasshoppers, and fish. I raised young jackdaws (little members of the crow family) and a magpie that had fallen out of the nest. Most Saturdays I took my self-made fishing net and rode my bicycle to visit ditches where I'd catch salamanders, stickleback fish, glass eels, tadpoles, bitterlings, and so on. My goal was to keep them all alive. I ended up with a little zoo in a shed behind the house with fish tanks, multiplying mice, birds, and an adopted kitten. I had no dog, but a large neighbor dog became my friend and was often by my side. I liked the smell of animals as well as their company. I still do.

Where would such interests fall on the scale of socialization through play? Animals move, like cars, but they also require nurturing, like dolls. Since my family didn't push me in this direction and at most tolerated my obsession, I was essentially self-socializing: a seeming contradiction in terms. I dreamed of my animals and how I would set up my first aquarium, or where I'd release my young jackdaws. I moved inexorably toward becoming an animal lover, which laid the foundation for my current profession. Affection for animals is not a gendered issue by any means, as one finds it in both boys and girls, both men and women. Yet I don't recall ever agonizing over whether my interests were sufficiently masculine.

Sweden, a nation that officially promotes gender equality, once pressured a toy company to change its Christmas catalog so that it featured boys with a Barbie Dream House and girls with guns and action figures.[7] But when the Swedish psychologist Anders Nelson asked three- and five-year-old children to show him their toy collections, things turned out differently. Almost every child had his or her own room with a staggering average of 532 toys. After going through 152 rooms and classifying thousands of toys, Nelson concluded that the collections reflected exactly the same stereotypes as in other countries. The boys had more tools, vehicles, and games, and the girls had more household items, caregiving devices, and outfits. Their preferences had proved immune to the equality ethos of Swedish society. Studies in other countries confirm that the attitudes of parents have little or no impact on children's toy preferences.[8]

Boys will make toy guns out of nothing, turn dolls into pulverizing weapons, transform a dollhouse into a parking garage, and make pots and pans (given as a kitchen set) move across the carpet like cars while producing *vroom! vroom!* noises. Boys are such noisy players! They love making loud vehicular and shooting sounds of a kind that one almost never hears from playing girls. I personally know a little boy whose first word was not *dadda* or *mamma* but *truck.* He later spontaneously began to call his grandparents by the brand of car they drove.

Play cannot be dictated. Give a girl a toy train, and she may rock it to sleep or put it in a baby carriage and cover it with a blanket before wheeling it around. It's the same as with our pets. We bring them fancy new toys, but they prefer to chew up an old shoe (if we're lucky) or chase a cork we accidentally drop onto the kitchen floor.

The American science writer Deborah Blum silently despaired at the stubborn tendency of the young to play however they like:

> *My son Marcus passionately covets toy weaponry. Denied even so much as one lousy plastic pistol by his gun-intolerant mother, he has compensated by building armaments out of everything from clay to kitchen utensils. I watched him charge after the cat, rushing about the house, shouting, "Shoot him with a toothbrush!" and I found myself mentally throwing up my hands.*[9]

We have three main ways of finding out whether human preferences have a biological origin. The first is to compare ourselves with other primates that lack our cultural biases, which is all of them. The second is to look at a large number of human cultures to see which preferences are universal. And the third is to test children so early in life that culture can't yet have influenced them.

Given my background, I prefer the first method. Considering the above experiments on toy preferences, one may wonder if the same tendencies are found in primates free from human influence. The primatologists Sonya Kahlenberg and Richard Wrangham report behavior in wild chimpanzees that's reminiscent of Amber with her broom. During fourteen years of fieldwork in Kibale National Park in Uganda, they documented many occasions of young chimps holding on to rocks or wooden logs in ways that looked as if they were carrying an infant. This behavior was three to four times more common in young females than in males. They might put their pet rock aside while foraging for fruits, only to pick it up again before traveling to another place. Sometimes they held the log or rock close while sleeping in their nest or even built a nest especially for it. Females played tenderly with these items as if they were handling an infant, whereas young males were less caring, and sometimes kicked a rock in the same rough way they kick each other. This behavior did not reflect imitation of mothers, because mothers never carried logs or rocks. The young females themselves gave up doing so as soon as they had their first baby.[10]

In Guinea, the eight-year-old (prepubertal) chimpanzee sister of a seriously sick infant followed her mother around through the jungle. The Japanese primatologist Tetsuro Matsuzawa said that to his surprise, the worried mother once "extended her arm to touch the infant's forehead. It looked as if she measured the fever." After her infant died, the mother wouldn't let go of the corpse and carried it around for days until it turned into a dried-out mummy. She swatted away the flies swarming around it. Perhaps sympathizing with her mother's tragic situation, the daughter developed the habit of carrying a short rod on her shoulders or under her arm like an infant. One time she put it down and "slapped the rod by one hand several times, just like softly slapping the back of an infant." Matsuzawa interpreted the young female's behavior as pretend-mothering. He compared it to the Manon people in the nearby village of Bossou, where girls imitate mothers with newborns by walking around with a stick doll attached to their backs.[11]

The latter observation relates to the second way to determine if human

A chimpanzee walks around in a sanctuary holding a doll on her back like a mother ape carrying her infant. Young female apes are drawn to dolls and in the wild practice maternal skills on wooden logs.

preferences are biological: look at a great variety of cultures to see which ones are universal. Are they found in all of humanity? Unfortunately, we have little cross-cultural information about child behavior. There are quite a few studies in industrialized societies, but obviously we'd want a broader range of cultures. The only study that covered a diverse cultural mix found that newborns appeal much more to girls than to boys. Girls typically help out caring for younger siblings. They do so under their mothers' watchful eyes, while boys often play away from home.[12]

Even the 1949 book *Male and Female* by the most celebrated anthropologist of the previous century, Margaret Mead, says remarkably little about child play. Mead interviewed twenty-five adolescent girls—no boys—from various Pacific island cultures. Toys didn't figure into her account. For Mead, the source of socialization was not children's play but the way adults talk about men, women, and their interactions in real life.

Mead's work is ground zero for gender socialization theory because she demonstrated how variable sex roles can be. It has inspired claims that these roles are mostly or entirely cultural. After rereading *Male and Female,* however, I am no longer convinced that this was Mead's main message. She discusses several worldwide truths about being male or female. For example, she claims that girls are always kept closer to home and permanently clothed, whereas boys of the same age may go about naked and are given freedom to roam. A boy also learns that he'll have a long way to go before he will ever be "the man who can win and keep a woman in a world filled with other men." Mead stresses the universality of male competition, stating that "in every known human society, the male's need for achievement can be recognized." Men, to feel fulfilled and successful, need to excel at something—to be better at it than other men and better than women.[13]

Every civilization needs to offer men opportunities to realize their potential. A recent survey of seventy different countries confirmed this difference. Universally, men put more value on independence, self-enhancement, and status, whereas women emphasize the well-being and security of their inner circle as well as people in general.[14]

To feel accomplished, women always have their biological potential to give birth. It's the one thing they can do that men can't. A mother's job is so vital to society and so fulfilling that Mead thought that men must resent their inability to match it. She coined the phrase "womb envy" as a counterpunch to Sigmund Freud's "penis envy." Later in life, Mead regretted her one-sided emphasis on culture. In the preface to the 1962 edition of her book, she noted, "I would, if I were writing it today, lay more emphasis on man's specific biological inheritance from earlier human forms."[15]

This brings us to the third way of gauging the role of biology. Shortly after a human child is born, we have a window of time to catch them before they know anything about gender or our hang-ups in this regard. When one-year-old boys and girls watched videos of moving cars and talking faces, the boys looked more at the former and the girls more at the latter. But since these babies may already have been influenced by toy culture, a follow-up study looked at infants at the earliest possible age. It tested one-day-old neonates in the maternity ward of an English hospital right next to their exhausted mother. The babies saw either the experimenter's face or a similarly colored object that was not a face. Coders who were blind to the babies' sex noted that females looked more at the face and males more at the object, suggesting that from day one girls are more socially oriented.[16]

Toy preferences, too, appear so early in life and are so pervasive that a recent review covering 787 boys and 813 girls from mostly Western cultures concluded: "Despite methodological variation in the choice and number of toys offered, context of testing, and age of child, the consistency in finding sex differences in children's preferences for toys typed to their own gender indicates the strength of this phenomenon and the likelihood that it has a biological origin."[17]

Color is a different issue altogether. Testing eighteen-month-old infants on a variety of pictures, boys looked more at cars and girls more at dolls, but the color of the pictures had no effect. The children showed no preference for pink or blue. Young children are not yet under the spell of the color coding that's all around us. The distinction between blue for boys and pink for girls was made up by the clothing and toy industries. At one time these colors were even reversed. Initially, all infants wore white, which was easier to clean and bleach. A 1918 article in *Earnshaw's Infants' Department* introduced the first pastel colors, saying, "The generally accepted rule is pink for the boys, and blue for the girls. The reason is that pink, being a more decided and stronger color, is more suitable for the boy, while blue, which is more delicate and dainty, is prettier for the girl." It is only relatively recently that the West settled on the reversed color binary. If these colors now appeal to children—with girls refusing blue and boys refusing pink and parents worrying about "perverting" their children by dressing them in the "wrong" color—this is purely a cultural choice.[18]

At the very least, there is much better evidence that culture affects preferences for colors than preferences for toys.

Focusing on toys and colors, however, risks overlooking the most dramatic sex difference of all when it comes to play. Found in a great variety of human cultures and in all primate studies, young males have an elevated level of energy and are more physically rambunctious than females of the same age.[19] That boys are three times more likely than girls to be diagnosed with attention deficit hyperactivity disorder (ADHD) reflects the same sex difference.[20] When children are free to play alone in a room, boys typically engage in unrestrained roughhousing, whereas girls have less body contact and tend to structure their play into a storyline.[21]

Scientists equipped 375 typical American boys and girls with accelerometers, a small device worn on the hip that measures body movements. Having done so for one week per child, they found that boys of all ages are consistently more physically active than girls. In terms of general activity, the differences were not marked, but girls showed far fewer bursts of vigorous motion than boys.[22] A similar study of 686 European children yielded the same outcome.[23] A review of more than one hundred different countries concluded that the greater physical mobility of boys is universal.[24]

I am always astonished at the inexhaustible energy with which young male apes romp around, jump up and down things, and go at each other, rolling over the ground with big laughing faces while they rip each other apart. Known as rough-and-tumble play, it's mostly fake assaults, wrestling, pushing, shoving, slapping, and gnawing on each other's limbs while laughing. Apes have open-mouth laugh faces and produce hoarse laughlike sounds that serve to make their intentions clear. This is essential to avoid confusion because social play often looks like fighting. If a young chimp jumps on top of another and puts his teeth in the other's neck while laughing, the other knows it's just for fun. If the same were to happen silently, it could be an attack, which obviously would require a different response. The laughter of chimps is so loud and infectious that when I hear it in my office at the Yerkes Field Station, which overlooks a grassy outdoor area with twenty-five apes, I often chuckle to myself at the fun they seem to be having.

There is much less rough-and-tumble play among females. Female chimpanzees wrestle too, but more languidly so that it rarely looks like a test of strength. They prefer different games, sometimes quite inventive ones. For example, two prepubertal females developed the habit of trying to reach my office. For a while, they played this game every day. First, they'd together move a large plastic drum right underneath my window. Then they'd settle on top of it, with one of them climbing on top of the other. The one below would start flexing and stretching her legs, up and down, like a springboard. The one standing on her shoulders would try to reach my window with her hands but never succeeded. Their cooperative venture was quite different from the mock fights among the males.

The males' exuberant boisterousness and displays of vigor explain why young females keep their distance. It's not the way they like to play. This is no doubt why sex segregation marks the play of all primates. Males generally play with males and females with females. Their interaction styles are more compatible, and females often retreat from male play initiations.[25] They do so without any of the gender instruction that takes place in our societies. In humans, too, sex-segregated play is the rule. Children all over the world create separate play spheres: one for boys, one for girls.[26]

For six months, Carol Martin and Richard Fabes observed sixty-one four-year-old American children during unstructured play and concluded:

The more boys played with other boys, the more positive emotions they were observed to express over time. Thus, although play among boys is rough and dominance oriented, boys appear to find this active type of play increasingly interesting and compelling. . . . Other research suggests that boys respond with aroused interest and a matching response when another boy makes a bid to initiate rough play, whereas girls do not.[27]

Not all schoolteachers like the wild play of boys, which they find too aggressive. This may be one reason why boys are disproportionally disciplined and expelled from school.[28] Most play among boys has little to do with aggression, though. This is easy to see from their facial expressions, the laughter, and the reversibility of roles (first one is on top, then the other), and especially from how they split. After having wrestled, boys part ways as happy friends.

Rough-and-tumble play serves male bonding and teaches crucial skills. Since in almost all primates, adult males are physically stronger than females and are more prone to confrontation, they must learn physical restraint at a young age. An adult male gorilla is so incredibly powerful that with just a little pressure of his knuckles on the chest of a baby, he could squeeze all air out of it. Yet silverback males do play with gorilla babies, and the babies survive. Males are so gentle that the mother will just sit nearby watching without any sign of apprehension.

Don't think that these inhibitions come naturally to animals: they are acquired. Over his long life, the big male has learned to check his movements while playing with weaker partners. This caution is known as *self-handicapping,* a phenomenon found in many animals, from a big dog play-fighting with a tiny one to a polar bear in the Arctic playing with a leashed sled dog, which he could also eat.[29]

The upper-body strength of men and women is so different that there is barely any overlap. Only a small minority of women even come close to the average physical strength of men.[30] It would be catastrophic, therefore, if men around the house were unconscious of their physical advantage. Fathers often play roughly with their children by throwing them up into the air before catching them, tickling them, or rolling around with them on the floor. Sometimes they let them have the upper hand. The shrieks of laughter tell us that children love these games and the risk and challenges they entail. Wrestling is particularly common between fathers and sons. As a result, children often look at Mom and Dad quite differently, turning to the former when they are upset and to the latter for play. As one review summarized, "Mothers' interactions with their children are dominated by caretaking, whereas fathers are behaviorally defined as playmates."[31]

The rough games of fathers teach children crucial first-hand lessons about men's strength while boosting their physical skills and self-confidence. But this works only with an extremely inhibited father who learned restraint during thousands of play bouts when he was a boy or young man. Wrestling games are a crucial part of socialization by fathers and male peers.

Roughhousing with young male chimpanzees taught me firsthand how these inhibitions are acquired. I would often give the apes a break from the intelligence tests that I was supposed to carry out as a student. Designed by a rat expert for whom all animals were simple learning machines, the tests were awfully repetitive and boring, far below the mental grade of a chimp. The two apes kept gesturing to me, urging me to join them for a play session. This was much more fun, also for me, but it became soon apparent that they were way too strong. These chimps had not even reached puberty yet: they were only four and five years old. If I hit them with full force on their backs, they'd just keep laughing, as if this were the funniest thing I'd ever done.

If they did the same to me, however, or took me in one of their impossible holds using both hands and grasping feet, I'd be in real trouble and have to protest ("ouch! ouch!"). They'd immediately release me and come around to take a close look at my expression with a concerned look on their faces to see what might be the matter. Who'd have thought that humans were such weaklings? If they saw that I was ready to resume play, we'd do so a bit more calmly. This is also how they regulate play among themselves and make sure that everyone is comfortable. The goal of rough-and-tumble is to have fun, not to inflict pain.

If one resists this process and tries to seek dominance, things may turn ugly. This happened to my successor, who continued the experiments on the two chimps after I left. Instead of dressing down, he arrived the first day wearing a suit and tie. He was sure he could handle such relatively small animals, mentioning how good he was with dogs. He must have tried to bully the apes in the playroom, though, not knowing that chimps always push back and have more strength in one arm than we have in all four limbs together. I still remember this student staggering out of the testing room, having trouble shedding the two chimps clinging to his legs. His jacket was in tatters, with both sleeves torn off. He was fortunate that they never discovered the choking function of a tie.

The play of girls and female primates is generally more nurturant, which is typically explained as an expression of the maternal instinct. I am skeptical about such framing, though, because the term *instinct* implies stereotypical behavior. "Instinctual" behavior sounds inflexible, not worthy of attention, because surely it requires no brainpower. The term *instinct* has fallen out of favor in the study of animal behavior. Even though all animals have inborn tendencies, just like humans, these are supplemented by lots of experience. This is as true for a natural activity such as flight (young birds can be incredibly clumsy while learning to take off and land) as for hunting, nest-building, and indeed mothering. Very few behaviors are instinctive in the sense of requiring no practice.

Among primates, the orientation to vulnerable newborns and their substitutes, such as dolls or logs, is undoubtedly part of biology and more typical of females than males. This is also true for dogs, for example. Pregnant or pseudo-pregnant dogs may gather up all the plush toys at home to guard and clean them. Attraction to baby or pup substitutes is logical given over 200 million years of mammalian evolution in which caring for offspring was obligatory for females and optional for males.

This is not to say, however, that females are born with maternal skills. A newborn infant may automatically root for a nipple, but the

A girl tenderly holds and kisses her newborn baby sister.
Girls' attraction to infants is a human universal.

mother still needs to learn how to nurse. This holds true for humans as
well as apes. Many apes fail to take care of their offspring at zoos due to
a lack of experience and examples. They don't hold their infant in the
right position for nursing, or they pull back if the infant latches onto a
nipple. They often need human models to fill the knowledge gap. Zoos
with a pregnant ape commonly invite women volunteers to demonstrate
how to feed a baby. Motherhood and bodily similarity naturally brings
humans and apes together. The ape observes the nursing human mother
and copies her every move once her own baby arrives.[32]

Young primate females are besotted with infants. They show far
more interest in them than do males.[33] Young females surround a new
mother and try to get close to her infant. They groom the mother and—
if they are lucky—get to touch and inspect her infant. Young males are

rarely present in this crowd, while females follow the mother wherever she goes. They may play with the newborn and carry it if the mother lets them, which serves as a preparation for the moment they get their own progeny.[34] Amber, for example, was a popular auntie for all infants in the chimpanzee colony. She carried, tickled, and held them and brought them back for nursing as soon as they became fussy. As a result, mothers were at ease if Amber solicited their babies, whereas they might be reluctant with other young females. They always turned down young males, who could be so rough and uncaring that they posed a danger. For example, a young male might take an infant up into a tree, which is a big no-no for every mother. Amber never did that.

The training of young females helps them later raise their own offspring by nursing, protecting, and transporting them. Motherhood is one of the most complex tasks a primate will face in life. When Amber had her first infant, she turned out to be a perfect mother right away. This is rare in apes, but it didn't surprise us.

Practicing maternal behavior, however, is by no means all that young primate females are interested in. In humans, dolls may find different purposes. U.S. presidential candidate Elizabeth Warren twittered out a picture of herself as a girl with a large set of dolls, writing, "I wanted to be a teacher since 2nd grade. Here I am with my doll collection—I used to line them up and play school."[35]

Primate females love playing imaginative games. In fact, one game became legendary in scientific circles as it hinted at make-believe by an ape. Until then, make-believe had been considered a uniquely human capacity. A first hint that apes are capable of pretense is when they, as we have seen, turn inanimate objects into made-up babies. But this particular case went further because the object was entirely fictional. It concerned Viki, a young chimpanzee, raised in the Florida home of Cathy Hayes.

In her 1951 memoir, Hayes included a chapter entitled "The Very Strange Case of the Imaginary Pulltoy." One day Hayes noticed Viki tracing a finger around the edge of a toilet bowl. At first, it looked as if she were carefully inspecting a crack in the bowl, but why would she be so intrigued? Then Hayes noticed that Viki seemed engaged in a tug of war, pulling strenuously at something invisible. Eventually, she gave a little jerk and pulled the "thing" toward herself, hand over hand, precisely the way she had previously done with toys on a string. To Hayes, it seemed that Viki had an invisible toy attached to an invisible rope that had gotten caught around the toilet.

In the days that followed, Viki played her game more often, confirming Hayes's suspicion. For instance, she would transfer the invisible rope from one hand to the other while looking behind herself, with one arm extended behind her to pull the toy. Once Viki called out in distress to her human mother when the imaginary rope had gotten stuck and she was unable to free it. She kept tugging at it while looking at Hayes. Hayes got in on the game and carefully disentangled the rope for Viki, who immediately tore off, dragging her invisible toy behind her.[36]

Hayes could hardly believe her own audacious interpretation, saying that she told her story just as a "bewildered mother." There is so much we don't know about the games of young primates. We always overlook the little ones. The play behavior of children, too, is grossly understudied. Even though children enthusiastically devote many hours a day to play, psychologists largely ignore it, while parents entertain the illusion that they are the architects of it. This is why we debate toys so intensely. The idea is that children have hardly any interests of their own and that we need to assist them by giving them gendered toys to mold them into "real" women and men. Alternatively, we steer them toward opposite-gender toys to allow them to grow into enlightened liberals. Both approaches are arrogant.

The best strategy would be to abolish all the typical divisions found in toy stores and accept the choices children themselves come up with, regardless of whether they fit our hopes and dreams. Step back and let them play in whatever way they like. Moreover, a great deal of play has little to do with toys or gender, such as my early fascination with animals and children's attraction to music, reading, camping, or collecting small objects, such as shells and rocks.

The only problem is that girls' clothing still doesn't include pockets!

2

GENDER

Identity and Self-Socialization

A cappuccino was all I wanted, one morning in 1991 at an international conference in Amsterdam. Standing in the hall of the convention center, holding my cup of java, I glanced up at a television screen. To my surprise, it featured a close-up of an erect human penis being stroked and licked. This wasn't pornography but rather an ad by a sex-therapy vendor. I noticed similar erotic scenes on other monitors. At this time of day, I expected the morning news! The city of Rembrandt and Anne Frank was an obvious choice for the World Congress of Sexology. Amsterdam has a famous red-light district, a huge annual gay pride festival, and the world's first sex museum.

Even if sexology is not my area, one can't study bonobos and not delve into it. Conversely, sexologists urgently need to hear about other animals. They are utterly human-focused—as if our species invented sex. Part of the problem is sexology's misconception that only humans enjoy recreational erotic activity. For other animals, it says, sex is purely procreational. I came to the conference to give a lecture on bonobos and disabuse sexologists of this bizarre notion. Most bonobo sex has little to do with breeding. They often do it in combinations unable to reproduce, such as between members of the same sex. They also have sex when they are still too young to reproduce, or while one of them is already pregnant. Bonobos have social reasons for sex. They're pleasure-seekers.

But enough about bonobos. While I was putting my slides in order (the vintage 35mm type), an older man in a crumpled gray suit strode with hasty steps into the hall. He might have been unremarkable save for his self-confidence and entourage. Like groupies surrounding a pop star, a dozen fawning young women and men stayed close to him wherever he went. They clamored to talk with him, take his coat, or fetch him a drink. I soon learned the identity of this man, who ignored his fan club. He was John Money, one of sexology's founders. Later in the day he would give a lecture entitled "Epidemic Antisexuality: From Onanism to Satanism."

In 1991 Money, a New Zealander–American psychologist, was at the peak of his fame. He was seventy and had given the world the vocabulary to talk more intelligently and kindly about sexual orientation, about being transgender, about atypical genital anatomy, about sexual identity, and indeed about gender itself. Before Money came along, those who failed to fit society's pigeonholes were customarily dismissed as deviants and freaks. It was this sexologist who in 1955 introduced the label *gender,* which until then had been used only for grammatical classification. In English, we recognize the gender of words such as *king* and *queen* or *ram* and *ewe.* In some other languages, the gender of nouns is reflected in articles, such as *le* and *la* in French, or *der* and *die* in German. Money borrowed this grammatical label, saying that for him gender refers to "all those things that a person says or does to disclose himself or herself as having the status of boy or man, girl or woman, respectively." He set gender apart from biological sex, aware of the occasional disparity between those two. He also founded the world's first Gender Identity Clinic at Johns Hopkins University in 1965. The terminology invented by Money gained immense popularity when feminism declared gender to be a social construct and when transgender people gained public recognition.[1]

I never saw Money again, but in later years, his entrances at conferences would no doubt have been less glorious. Despite all his accomplishments and his widely read books, he lost his reputation. His downfall came from an underestimation of biology. He got involved in the sex reassignment of a Canadian boy who lost most of his penis in a botched circumcision. Money (the scientist) persuaded the boy's parents to remove his testicles as well and raise him as a girl. His birth name was Bruce, but he became Brenda. Brenda wasn't told about his original sex.

Having made regular visits to follow Brenda's progress, the sexologist claimed unmitigated success. He triumphantly declared gender to be purely a matter of upbringing. Until a certain age, you could change a boy into a girl, and vice versa. Many people welcomed this news because it suggested that we have control over our destiny. Money became a hero to the women's movement. In 1973 *Time* praised his work as providing "strong support for a major contention of women's liberationists: that conventional patterns of masculine and feminine behavior can be altered."[2]

It all fell apart in such a dreadful way that Money became a controversial figure. Years after his death, he's still considered by some a charlatan and a fraud. The boy who was supposed to have become a girl fiercely resisted his new gender. Brenda was clothed like a girl and given dolls to play with, yet he walked and talked like a boy, tore off his frilly dresses, and stole his brother's trucks. He wanted to play with boys, build forts, and engage in snowball fights.[3]

Lacking a penis, he had been taught to sit down on the toilet. Nevertheless, he felt an irrepressible urge to urinate standing up. This caused friction with his classmates at school. Girls called him "cave woman" and banned him from their bathroom. Boys did the same—since he was dressed like a girl—so he ended up urinating in a back alley.

Only at the age of fourteen did Brenda finally learn the truth. It came as a relief since it explained so much, including why he had felt miserable for so many years. Under the new name of David, he returned to the identity of his birth. Tragically, he committed suicide at thirty-eight.

This heartbreaking story—known as the David Reimer case—contains an important lesson for those who believe biology can be ignored. In his zeal to present an optimistic picture, Money had downplayed signs of trouble. In the end, his intervention proved the exact opposite of what he had wanted to show. It made clear that a surgery followed by years of estrogen treatment and intense socialization still can't overturn a boy's male identity. Since then we have gained a better understanding of the interplay between nature and nurture and know that it's more complex than either Money or his detractors thought. But thanks to Money, we have at least a vocabulary to talk about it.[4]

The term *gender* has become an indispensable part of the discourse, even though it's getting overused. This is due to the failure of English to distinguish between sex and sex. "Having sex" employs the same word as "being of a particular sex." This confusion doesn't exist in every language, but explains why in American English, *gender* has begun to fill the void. The word has overtaken *sex* even when the latter term is more appropriate. At the zoo, for example, people will ask, "What is the gender of that giraffe?" In scientific journals, we may see titles such as "Sexual differences as adaptation to the different gender roles in the frog." A website on canines explains, "Identifying a puppy's gender is important: You do not want to end up with a dog of the sex you didn't want."[5]

Strictly speaking, this usage is incorrect. If the term *gender* refers to the cultural side of an individual's sex, its use should be limited to individuals who are affected by cultural norms. Despite the evidence for animal culture, I'd rather assign giraffes, frogs, and puppies a sex than a gender. Even the "gender reveal parties" thrown for human pregnancies shouldn't be called that because the unborn haven't yet been exposed to culture. They have no gender, only a sex.

It's hard to resist the new *gender* usage, though. As you will notice, I sometimes fall for the convenience myself. Ironically, a term put forward as an alternative to biological sex has come to stand for it. This obviously muddles the discussion of a delicate topic.

M OST OF THE TIME, the term *gender* covers culturally assigned roles, such as in the following definition by the World Health Organization: "The characteristics of women, men, girls and boys that are socially constructed. This includes norms, behaviors and roles associated with being a woman, man, girl, or boy, as well as relationships with each other."[6]

Gender is like a cultural coat that the sexes walk around in. It relates to our expectations of women and men, which vary from society to society and change through the ages. Some definitions are more radical, though, in that they seek to denaturalize gender. In those definitions, gender is an arbitrary construct quite separate from biological sex. The coat walks on its own, so to speak, and its styling is up to us.

The first version of the gender concept is uncontroversial. In our daily lives, we can easily see how society molds gender roles and pressures everyone to fall in line. The more radical notion of gender, on the other hand, clashes with what is known about our species' biology. Whereas it is true that gender goes beyond biology, it's not created out of thin air. The whole reason we have a gender duality is that the majority of people can be divided into two sexes. This doesn't mean that we should buy into everything associated with gender, such as the power imbalance between men and women. It also doesn't mean that we need to limit ourselves to two genders. But there are certain core elements that we're born with. As Money found out, this includes gender identity.[7]

Gender is one of the first things we pick up on when we meet a person. It's a crucial piece of information about anyone we wish to interact with. In experiments, people viewing a portrait in which all hair has been cropped out need only a second to guess the person's gender with nearly 100 percent accuracy.[8] In real life, identifying gender is often helped by cultural overlays, such as the way we dress, style our hair, splay or cross our legs, or bring a cup of tea to our lips. This is how we signal gender to the rest of the world. The importance attached to these signals explains why they are closely monitored. Women who spit on the ground or burp loudly are told that they are not ladylike, whereas men often get away with such behavior. The superstructure of gendered customs can be arbitrary to the point of being trivial. Moreover, it's far from stable over time. Thus seventeenth-century French male nobility walked around perfumed, in high heels, wearing embroidered clothing and long-haired wigs.

Other gender norms are more consequential, such as the education and jobs that are favored for either men or women. Insofar as these norms restrict choice, especially for women, they are rightfully under attack. The most meaningful expressions of gender have deeper roots, including the generally greater physical combativeness of men or the devotion of many women to children. These expressions are human universals that we share with other primates. Female care for the young is a mammalian trait.

Every human tendency, regardless of whether we rate it as natural, can be amplified, weakened, or modified by culture. Thus male aggressiveness may be glorified in one place and time, such as in a nation at war. In another place and time, however, it may be curbed to the point that open conflict is rare, and murder is almost unheard of.[9] Nevertheless, we shouldn't let the influence of culture lure us into thinking that the human aggressive instinct is a myth. The most common error in nature vs. nurture debates is to take proof of one influence as evidence against the other. If the gallons of ink spilled on the biological basis of altruism, warfare, homosexuality, and intelligence have taught us anything, it's that every human trait reflects an interplay between genes and environment.

A good example is language. Our mother tongue may seem purely cultural. A baby born in China will learn Mandarin, and one born in Spain will learn Spanish. We know from international adoptions that this has nothing to do with genes. Switched at birth, the first baby will learn Spanish and the second Mandarin.

Nevertheless, had these infants belonged to another primate species, they'd never have uttered a word. Science has known no shortage of attempts to teach language to our fellow apes, but the results have been disappointing. The human language faculty is unique and biological. We even know some of the genes involved. Our brains evolved to suck up linguistic information during the first few years of life. This means that we have both nature and nurture to thank for the language we speak.[10]

This combination, which is typical of biological processes, is known as a *learning predisposition*. Many organisms need to learn certain things at a particular time in their lives and are programmed to do so. In the same way that we are prepared to acquire language while young, ducklings imprint on the first moving object they meet. Sometimes, as in the case of Konrad Lorenz, this may be a bearded, pipe-smoking zoologist. The birds will follow this "parent" on walks and swims. This is not how it is supposed to go, however. Under natural conditions, ducklings dribble and paddle single file behind Mom. For the rest of their lives, they'll identify with the species she belongs to, which also happens to be their species. This is the whole point of imprinting.

Human gender roles are subject to similar learning predispositions. These roles themselves are not necessarily biological, certainly not in all their details. They are culturally acquired, but with a speed, eagerness, and thoroughness that's astonishing. The ease with which children adopt them hints at a biologically driven process. They imprint, as it were, on their gender the way ducklings imprint on their species. Children typically love to emulate adults of their gender whether they are real or fictional. Affected by the media, girls dress up like fairy-tale princesses, and boys slay dragons with swords. Children thoroughly enjoy these reenactments. Neuroimaging studies indicate that imitating people of one's own gender activates reward centers in the brain, whereas imitating people of the opposite gender does not. This doesn't necessarily mean that the brain is in charge, because it too reacts to the environment. But it does suggest that evolution has equipped our young with a feel-good bias to conform to their gender.[11]

In one early study, toddlers watched a short film in which a man and woman performed simple activities, such as playing an instrument or building a fire. The actors did so at the same time but on opposite ends of the screen. The children zoomed in on the actor of their gender: the woman was watched more by the girls than by the boys, who focused more on the man. The investigators interpreted this own-gender preference as follows: "It becomes increasingly relevant for them to learn and to adopt the social rules concerning male-appropriate and female-appropriate behaviors."[12]

We tend to view socialization as a one-way street in which parents teach their children how to behave, but *self-socialization* is at least as important. Children themselves seek and enact it. Fascination with persons of their own gender makes them pay attention to the behavior they wish to emulate. Here is how the American anthropologist Carolyn Edwards, inspired by her observations of boys and girls in a wide range of cultures, defined self-socialization: "The process whereby children influence the direction and outcomes of their development through selective attention, imitation, and participation in particular activities and modalities of interaction that function as key contexts of socialization."[13]

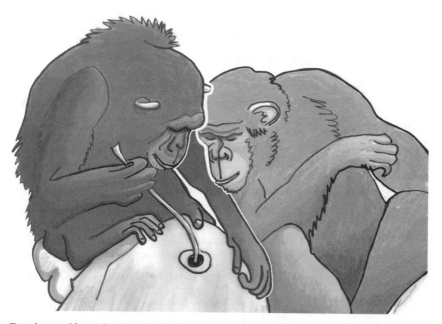

Daughters self-socialize by adopting their mother as role model. A young female chimpanzee (right) watches closely how her mother fishes for termites.

Sᴇʟꜰ-ꜱᴏᴄɪᴀʟɪᴢᴀᴛɪᴏɴ ᴀᴘᴘʟɪᴇꜱ ᴛᴏ other primates as well. In the African rainforest, young chimpanzees learn from their mother how to extract termites by dipping twigs into the insects' nests. Chimpanzee daughters faithfully imitate the specific fishing technique of their mother, but sons do not. Although both spend equal time with their mom, the daughters watch her more intently during termite feeding. Mothers also share their tools more readily with daughters. This way young females learn what the right tool looks like, whereas young males fend for themselves. For sons, the maternal example may be less relevant, because later in life they will derive most animal protein from hunting monkeys and other large prey.[14]

A similar learning bias occurs in wild orangutans. By the age of eight, close to adolescence, daughters eat the same foods as their mother, whereas sons have a more diverse diet. Having paid attention to a wider range of models, including adult males, young males even consume foods that their mother never touches.[15]

Among wild capuchin monkeys in Costa Rica, the young need to learn how to open *luehea* fruits. These fruits are full of nutritional seeds, which are removed either by vigorously pounding the *luehea* fruit or by roughly scrubbing it on a branch. Every adult female employs one or the other technique, which her daughters copy. For the rest of their lives, daughters will either pound or scrub like their moms. Sons, in contrast, are unaffected by their mother's example.[16]

From studies on social conformism in other primates, we know that individuals acquire habits from those to whom they feel close. Observational learning is guided by bonding and identification.[17] Daughters not only copy their mother's eating habits but also learn from her how to raise infants. The role models of young males are harder to pinpoint as they usually don't have a clearly defined father figure. Their father being unknown, they follow the example of adult males in general. Thus, wild vervet monkey females preferentially copy female models regardless of their effectiveness at opening a foraging box set up by scientists. Males, however, copy models of both sexes and especially successful males.[18]

Young males like to hang out and groom with older males. In Kibale National Park in Uganda, adolescent male chimpanzees develop special friendships with aging males. Between twelve and sixteen years old, adolescents are independent of their mothers but are not yet ready to fight their way into the adult male hierarchy. Like human teenagers, they are in between childhood and adulthood. Their favorite friends, apart from peers, are males of around forty. These males are over the hill and have mostly "retired" from power politics. The young and old make a great combination. The retired males are easygoing and pose no danger, which makes them ideal models. DNA analysis shows that older males often are the biological fathers of the adolescents who seek their company.[19]

Fascination with male models may start much earlier, though. Chimpanzee infants seem to follow the bluff displays of adult males very carefully. Each male has a distinctive style, which includes spectacular jumping, hand clapping, flinging objects around, breaking off branches, and so on. I knew one alpha male who had a habit of drumming for minutes against a specific metal door to accentuate his performances. The din he produced served to announce his vigor to the entire colony. While he drummed, females would keep their youngest offspring close to them, because males in this agitated state are unpredictable. Once the male calmed down, the mothers would let their little ones go. Often an infant male—never a female—would walk up to the exact same metal door. He'd have all his hair on end and kick the door just like alpha had done. It didn't sound the same, but he got the idea.

If the monkey-see-monkey-do mentality of primates encourages self-socialization by emulating same-sex models, the gender concept might apply to them as well. Behavioral differences between the sexes may be partly cultural. We need more studies than the handful mentioned above, but it's time to revisit the maxim that "every species has sex, only humans have gender."

ONCE UPON A TIME, science thought humans were endlessly flexible. This idea was especially popular among anthropologists, who traditionally emphasize culture at the expense of biology. In the 1970s Ashley Montagu described our species as utterly devoid of inborn tendencies, claiming, "The human being is entirely instinctless." Mind you, a decade before, the same Montagu had glorified women as intrinsically more loving and caring than men.[20] There is a blatant contradiction here. One cannot treat the human mind as a blank slate upon which culture carves gender norms while at the same time positing a natural difference between the sexes. This may explain why Melvin Konner, an anthropologist who agrees with Montagu on female superiority, distanced himself from the culture-is-everything mantra of his discipline:

> *Boys and girls really are different, and so are the men and women they become. It is a deep biological and philosophic insight, and although I did not at first accept it—I was a strong cultural determinist in my youth—I am glad to embrace and defend it now.*[21]

We have absolutely no need, however, to choose between culture and biology. The only plausible position is to be an *interactionist*. Interactionism assumes a dynamic interplay between genes and the environment. Genes by themselves are like seeds dropped onto the pavement: they can't produce anything on their own. Similarly, the environment by itself is hardly relevant because it requires an organism to act upon. The interplay between these two is so intricate that most of the time we are unable to disentangle their contributions.[22]

Hans Kummer, a Swiss primatologist, came up with a helpful analogy to explain why this is so. Asking if an observed behavior is due to nature or nurture, he said, is like asking whether the percussion sounds we hear in the distance are produced by a drummer or the drum. It's a silly question because, on their own, neither one makes any noise. Only if we were to hear distinct sounds on different occasions could we legitimately ask if the difference was owing to a change in the drummer or the instrument. Kummer concluded, "Only a difference in traits, not a trait as such, can be called innate or acquired."[23]

This insight came from someone who had wondered all his life about the origin of observed behavior. Interactionism is not very popular, though, because it provides no easy answers. The media often try to give us one ("this trait is 90 percent genetic"), but such statements are nonsense. Just as we can't specify the relative influence of the drummer and the drum, we can't specify the contribution of genes and environment to any given behavior. If a girl laughs exactly like her mother or a boy talks like his father, it may be because they flawlessly parrot their role models. But both children also inherited their parents' larynx and vocal timbre. Without a controlled experiment (and the ethical issues this would raise), we have little hope of untying the roles of genes and the environment.[24]

A similar problem arises for anyone who wants to know the origin of gender roles. Except for purely cultural embellishments—such as pink for girls and blue for boys—these roles integrate both nature and nurture. As a result, they are more resistant to change than you'd expect. In this day and age, some parents opt for a gender-neutral upbringing of their children to throw off what they see as society's shackles. They refuse to reveal their child's anatomy, sometimes not even informing the grandparents. They cut the hair of girls or let that of boys grow long, and they allow their children to dress any way they want, also if their son decides to go to school in a tutu. They do so in reaction to society's gender stereotyping and the associated inequality.

Note, however, that only one of the two words in *gender inequality* refers to a problem, and it's not *gender*. No one would propose to fight racism by urging people of different races to try to look more alike. So why would we try to get rid of gender? Ultimately, such attempts fail to address the deeper problem of inequality. It blames the existence of genders for the moral and political shortcomings of society.

For many, being a man or a woman is a source of pride and joy. People don't merely adopt gender identities but positively embrace them, regardless of whether we consider them cultural or not. We also shouldn't forget that, as the song goes, love is what makes the world go around. And aren't romantic love and sexual attraction intensely gendered for most of us? This is true whether we are attracted to members of the other sex or our own. I'm not sure, therefore, that raising children genderless does them much of a favor. How will they navigate the world and their feelings for others once puberty hits? Will their love life be gender neutral, too? I find this hard to imagine, even though I realize that a younger generation believes it possible.

Since donna was a tiny infant, she and I played together at the Yerkes Field Station. The little chimp would come running up to me whenever she saw me walk by. Turning around while pressing her back against the fence, she would look back at me over her shoulder. As soon as I'd dig my fingers into her neck and sides, she would giggle her hoarse chimpanzee laugh. Sitting in the distance, her mother Peony, grooming another female, would barely look up. Given Peony's extreme protectiveness, I took this as a compliment.

Later in life, Donna kept inviting me this way, even after she reached an age when most apes aren't ticklish anymore. She also frequently played with the big males in her group. The alpha male would seek her out for wrestling matches. Always gentle, he had a habit of roughhousing with juvenile males but not with females other than Donna. He could play with her for many minutes at a time, tickling and laughing as if she were the best playmate ever. This was the first sign that she differed from her sex peers.

Donna grew into a robust female who acted more masculine than other females. She had the large head with the roughhewn facial features typical of males, and sturdy hands and feet. She could sit poised like a male. If she raised her hair, which she did more often the older she got, she was quite intimidating, thanks to her broad shoulders. Her genitals were those of a female, however, even though they were never fully swollen. Female chimpanzees, at the peak of their thirty-five-day menstrual cycle, sport inflated genitals. But after Donna passed puberty, hers never reached the shiny maximum size that announces fertility. The males were barely interested in her and refrained from mating. Since Donna also never masturbated, she probably didn't have a strong sex drive. She never had offspring.[25]

Donna's menses were heavier than those of other females, with considerably more blood loss. Usually she was in good spirits, friendly and playful, but not during her period. We barely noticed the menstruation of other females and didn't see much of a mood change. Donna, by contrast, looked down and tired. It may have been due to pain or anemia. Since we noticed her mouth and tongue turning pale, we gave her iron supplements.

Curiously, most students of primate behavior barely talk about the sort of gender diversity illustrated by the case of Donna. There are always some males with less machismo than others, and always some females who act tomboyish. Those females enjoy rough wrestling more than others and initiate more daring games. Even though animal "personality" is a popular topic of study, science still ignores variability in relation to sex roles. It's perhaps as with our species, in which for the longest time we neglected exceptions to the binary rule. Here, too, the distinction between sex and gender comes in handy. I like the way Robert Martin, a British biological anthropologist at Chicago's Field Museum, put it. Most differences between the sexes are bimodal, he wrote, whereas differences between the genders move across a spectrum.[26]

Largely defined by chromosomes and genitals, sex is binary for the vast majority of humans. In the language of digital electronics, the word *binary* refers to the two-number system of ones and zeroes (1/0). Applied to sex, *binary* means that individuals are born either male or female. Exceptions with regard to both chromosomes and genitals do occur, however, making the sexual binary an approximation at best.[27]

Differences between the sexes are rarely black and white, however.

Rather, they show a bimodal distribution (the famous bell curves), which means that they concern averages with overlapping areas between them. For example, men are taller than women, but only in a statistical sense. We all know women who are taller than the average man, and men who are shorter than the average woman. The same overlap holds true for behavioral traits, such as when men and women are said to differ in assertiveness or tenderness.

Gender is a different issue altogether. It relates to culturally encouraged sex roles in society and the degree to which each individual expresses and fits them. For gender, the appropriate terms are not *female* and *male* but *feminine* and *masculine*. These terms refer to social attitudes and tendencies that aren't easily classified. They often mix so that aspects of both are manifested in a single personality. A man may be both manly and have a feminine side, and a feminine woman may occasionally express herself in a distinctly masculine fashion. Gender resists division into two neat categories and is best viewed as a spectrum that runs smoothly from feminine to masculine and all sorts of mixtures in between.

On this gender spectrum, Donna was much more on the masculine side than most members of her sex. Even her body hair reflected this. As in our species, the male chimpanzee is the hairier sex. This allows him to look larger than life when he "goes pilo" (from *piloerection,* or bristling hair). Donna had unusually long hair and could go pilo all over her body like a male. She furthermore often acted as if she were part of the male world. As soon as males started bluffing around, intimidating the group with their noisy hooting displays, Donna would join in and charge by their side. She'd sway her body or perform a "bipedal swagger." On two legs with her arms hanging loose and all her hair on end, she'd adopt a wide-legged gunslinger walk. Like the "rain dance" of wild chimpanzees, a sudden downpour could prompt her to walk around like this. You'd swear you were seeing a male.

Male displays rarely end in an attack. More commonly, their hooting culminates in a vocal climax, like a drawn-out rallying cry. Donna's cry was higher pitched than that of the males, but that she uttered one at all was atypical for a female. Acting like the sidekick of adult males, she might gain temporary dominance. She was only middle-ranking, but even females ranking above her would step out of her way when she entered this worked-up state.

The males themselves let Donna act out as if they didn't notice. If she'd been another male, they might not have done so—they watch their rivals closely during displays and provoke or react to them. But Donna posed no threat. She did not compete with them and was non-aggressive. Bluff and swagger don't count as aggression, as long as they don't escalate to charges or attacks. My team, having collected over a hundred thousand data points over the years by observing the colony from a tower, found Donna to be the least aggressive individual in our study. Her grooming and play behavior were comparable to that of other females, but she neither gave nor received much aggression. She managed to stay entirely out of trouble.

Donna was no pushover, though. She had the benefit of a dominant mother, who was always ready to jump in, and she also stood up for herself. Once a female didn't take well to her hooting and swaying and hit out at her screaming. Donna went after her and pummeled her back with her fists. Normally dominant, this female submitted to the abuse. Donna had acted in self-defense, though. She never did anything like this for no reason.

Before I wrote about Donna, I asked my co-workers what they thought of her. Some of my research team are gay or lesbian and told me they looked at this female through rainbow-colored glasses. They were all fascinated by her atypical behavior and remembered her fondly. None considered her lesbian, though, because she didn't seek sexual contact with other females. All thought she was well accepted despite her inclination to throw her weight around. It was just part of who she was, and neither the human observers nor the other apes seemed to mind. She had a happy-go-lucky attitude and got along with everyone.

Whether we might call Donna "trans" is beyond my ken, because with animals, this is impossible to know. Individuals who are born as one sex yet feel they belong to the opposite sex are known as *transgender*.[28] Transgender humans actually prefer to turn this description around and prioritize their felt identity. They were born as one sex but found themselves inside the body of the other. We have no way of applying this to Donna, however, because we can't know how she perceived her gender. In many ways—her grooming relations with others, her nonaggressiveness—she acted more like a female than a male. The best way to describe her is perhaps as a largely asexual gender-nonconforming individual.

Over decades of working with apes, I have known quite a few whose behavior was hard to classify as either masculine or feminine. Even though they form a minority, nearly every group seems to have one. There are always males, for example, who do not play the status game. They may be muscular giants yet retreat from confrontations. These males never reach the top, but they also don't sink to the bottom because they're perfectly capable of defending themselves. They are ignored by other males, who have given up recruiting them as allies for their political machinations. A male who is unwilling to take risks is no help in challenging the higher-ups. Females, too, have less interest in these males as they are unlikely to stand up for them whenever they are bothered by males or other females. For this reason, males without a dominance drive lead relatively quiet but isolated lives.

Unfortunately, we have no idea how common nonconforming individuals are, because scientists look for typical behavior. We like to form a clear picture of how females and males behave. We go for the peaks of the bimodal distribution while ignoring the valleys. Irregularities remain underreported.

When I last saw Donna, she was a young adult. As I was saying hello, she locked eyes with me, then jerked her head to stare at something in the grass on my side of the fence. This is the no-hands way a

TABLE: *Common vocabulary in relation to human sex and gender.*

Terminology	Definition
Sex	The biological sex of a person based on genital anatomy and sex chromosomes (XX for female, XY for male).*
Gender	The culturally circumscribed role and position of each sex in society.**
Gender role	The typical behavior, attitudes, and social functions of each sex resulting from an interplay between nature and nurture.
Gender identity	A person's inner sense of being either male or female.
Transgender	Referring to a person whose gender identity does not match their biological sex.***
Transsexual	Referring to a person who has undergone hormonal and/or surgical gender reassignment; a medical term.
Intersex	Referring to a person whose sex is ambiguous or intermediate since their anatomy, chromosomes, and/or hormonal profile doesn't fit the male/female binary.

* This is the medical definition of human sex. In biology, sex is defined by the size of gametes (such as sperm and eggs), with females having the larger gametes.
** In the United States, the term *gender* is increasingly used for biological sex, including even that of animals, which is not its original meaning.
*** When gender identity and biological sex agree, the person is said to be *cisgender.*

chimp points things out. I followed her stare to a stick she had spotted. As soon as I handed it to her, she ran off to join a "cooking" circle formed by her friends in the large enclosure. For a while, youngsters had a game of digging a hole in the ground and putting water into it. They'd sit around and poke the mud with sticks. We called it "cooking" because superficially it looked as if they were making a stew. One youngster would pick up a plastic bucket and walk to the water faucet to fill it up. He or she (both sexes played this game) would then slowly walk the long way back, careful not to spill any water. The bucket's contents would be dumped into the hole for a fresh round of stirring.

Young chimps are always inventing games, which they play for a few weeks until one of them comes up with a new one. Donna seemed too old for this, but she enjoyed the joint entertainment. A stocky female contentedly sitting among the little ones is how I remember her.

THE EXISTENCE OF transgender people challenges the notion of gender as an arbitrary social construct. Gender roles may be cultural products, but gender identity itself seems to arise from within.[29]

When people are asked how they identify, the number of transgender persons is relatively high. The latest estimate is that 0.6 percent of adults are transgender, which translates into 1 or 2 million people in the United States alone. But this number is almost certainly an underestimate.[30] Transgender persons have every reason to be reluctant to come forward. Remember the bathroom bills that tried to erase them from the public sphere? Currently, similar efforts are under way in relation to sports. Instead of accommodating transgender people and recognizing their rights, American society seems intent on demonizing them and making their lives difficult. The big mistake, which we have faced before with respect to homosexuality, is to present being transgender as a disorder that needs fixing or a choice that needs correcting, as if it were a mere lifestyle preference.

But being transgender is intrinsic and constitutional. By "constitutional," I mean the opposite of socially constructed. It is a trait that taps into the essence of who we are. We don't know if being transgender is caused by genes, hormones, experience in the womb, and/or early postnatal experience. What we do know is that it typically arises early in life and cannot be reversed. One of the best-known examples is Jan Morris, who opened her book *Conundrum* with the line: "I was three or perhaps four years old when I realized that I had been born into the wrong body and should really be a girl. I remember the moment well, and it is the earliest memory of my life."[31]

Gender socialization invariably takes genital anatomy as its starting point. Transgender children, however, resent the expectations imposed on them. Their socialization, instead of being a cooperative affair between parents and children, often turns into an angry war of rebellion and coercion. Devon Price, who was declared a girl at birth, gives a coming out narrative that illustrates both the absence of choice and a strong desire to emulate the gender he felt he belonged to:

> *People pushed female norms onto me, and I tended to either dismiss those norms, or fail at them. I thereafter received the socialization that typically follows a child failing to meet gendered norms. I was, to an extent, perceived and socialized as a gender fuck-up, not as a girl. I always knew on some level that I wasn't a cis girl, and I automatically dismissed some female gender norms that struck me as ill-fitting or unfair. I was always very very averse to expressing emotional pain or weakness. I always emulated men in terms of how authoritatively I spoke or expressed ideas. Throughout my life, I have wanted to be more like (stereotypical) men in terms of assertiveness and directness.*[32]

No one urges transgender children to accept who they are, at least not initially. On the contrary, parents, siblings, teachers, and peers get upset whenever a child takes on the appearance and habits of a different gender. They punish, mock, lecture, bully, and ostracize them. Despite this intense hostility, trans children stubbornly develop according to their felt identity, which goes to show that it's not the environment that constructs their gender. It's the child itself.

The largest study to date tested 317 American transgender boys and girls at an average age of seven and a half years.[33] They were compared with siblings and children whose gender was congruous with their assigned sex. In other words, transgender boys (boys born with female anatomy) were compared with cisgender boys (boys born with male anatomy), while transgender girls were compared with cisgender girls. Information was gathered on toy preferences (dolls vs. trucks), clothing style (dresses vs. pants), preferred playmates, and expectations about their future as men or women. The last information was striking, because transgender children were just as convinced of their future gender as cisgender children.

Transgender and cisgender children developed in almost the same way. A child who was born with male genitals and was raised for ten years as a boy, but who considers herself a girl, turns out to be just as feminine in her social attitudes, preferred toys, hairstyle, and desired clothing as her sister who was born a girl. The same applies to a child who was born with female genitals but considers himself a boy. This child will turn out just as masculine as his brother. The investigators concluded that "neither sex assignment at birth nor direct or indirect sex-specific socialization and expectations (such as rewarding masculine things and punishing feminine ones for assigned males) . . . necessarily define how a child later identifies or expresses their gender."[34]

A tiny area in the brain, known by the lengthy name *the bed nucleus of the stria terminalis,* seems involved in gender identity. It's one of the few brain regions that differs between the sexes, being about twice as large in men as in women. Dick Swaab's neuroscience institute in Amsterdam made the first postmortem dissections of transgender brains to inspect this specific area. It was found to be female-like in transgender women despite their natal sex having been male. In the brain of one transgender man, it looked male-like despite his natal sex having been female.

It seems, therefore, that the brain offers a better indicator of the gender that people claim for themselves than their genital anatomy. This doesn't mean that we have found the holy grail of gender identity, though. As the scientific mantra goes, correlation doesn't mean causation. It is hard to tell if the size of this brain region is the source or the product of gender identity.[35]

One speculation is that in a fraction of human pregnancies, the body takes off in a different direction than the brain. A fetus's genitals differentiate into male and female during the first few months of pregnancy, whereas the brain differentiates by gender in the second half of pregnancy. If these processes got disconnected, it could make the brain assume one gender while the body assumed another.[36]

Gender identities are probably shaped in the womb through hormonal exposure. Experience after birth seems to have little impact. This could explain why no amount of conversion therapy, combined with prayer and punishment, changes the minds of transgender persons. Therapies to "repair" or "cure" LGBTQ individuals are widely recognized as pseudoscience. They are as misguided as attempts to correct left-handedness. Not every human trait is malleable. Mental health organizations warn that these therapies do more harm than good and should be banned.

Gender identities that are congruous with the body, such as those of the majority of people, are hardly any different. We start life with a particular identity or develop one soon after birth. It is an essential part of ourselves that we flesh out through self-socialization. For most children, this identity fits their genital sex, whereas for trans children it's the opposite. They all know who they are and what they wish to become, and they seek out information that suits their identity and temperament. Joan Roughgarden, an American biologist who is transgender herself, envisions gender identity as a cognitive lens:

> *When a baby opens his or her eyes after birth and looks around, whom will the baby emulate and whom will he or she merely notice? Perhaps a male*

*baby will emulate his father or other men, perhaps not, and a female baby
her mother or other women, perhaps not. I imagine a lens exists in the brain
that controls who to focus on as a "tutor." Transgender identity then is the
acceptance of a tutor from the opposite sex.*[37]

We learned from John Money to distinguish culturally inspired
gender roles from biological sex. This dichotomy is front and center in
the ongoing debate about the changing positions of women and men
in society. At the same time, however, Money taught us that the two
are never fully disconnected. He may not have meant to make this
point, but this is the lesson derived from his claim of having turned a
boy into a girl. He never did. He viewed the child as a passive recep-
tacle of society's expectations, but the actual locus of control is the
child itself. The child in question was born with a gender identity that
urged him to self-socialize as a boy despite all the dresses and girl toys
thrown at him.

Instead of choosing between nature and nurture, self-socialization
combines the two. It comes from within but adopts the outside world as
its guide. It allows children to become whom they want to be.

3

SIX BOYS

Growing Up Sisterless in the Netherlands

The arrival of six boys in a row was deeply disappointing for my parents. After the first three, they were more than ready for a daughter. My mother had saved her own mother's name—Francisca—for the happy occasion. When I became son number four, she lost all hope and named me after the same saint instead. It turned out to be a perfect choice, because even though I lost my faith long ago, the one saint I find easy to admire is Francis, the patron saint of animals, whose feast day of October 4 is World Animal Day.

In those days, a child's sex remained unknown until birth. My father had calculated that the chance of getting a fourth son was less than 10 percent. The probability of having a boy, however, remains 51 percent for every consecutive conception. My parents must have stayed optimistic until the very last minute. After my birth, my mother went into a depression. She was lifted out of it, as she told me many times, only because I was such an upbeat child. Every time she picked me up, I'd cheer her up. She saw this as a deliberate trick on my part, as if I had decided that the only way to survive a dispirited mother would be to smile and coo all the time. My own theory is that I'm a born optimist.

Having grown up among so many boys, I am at ease around men. Perhaps too much so, because I don't share the odd prejudice according to which men are hard on each other and must live in constant stress. Once while some male colleagues and I were relaxing after a conference, we discussed this issue. One of them complained that men are always testing each other and trying to gain the upper hand. He was so upset by the way men undermine each other, at least in his mind, that he choked up! I couldn't believe how traumatized he was until he added that he had grown up as a single child. This background must have kept him from figuring out the paradox of male relations. On the surface, the power dynamic is real, which is why you should never insult or provoke a man for no reason. At the same time, however, it's also a game. Tests and insults are just the opening salvos. Soon afterward men move on to banter and jokes, and before you know it, we get comfortable, bonded even. This is how men relate to each other and check out who's worthy of their attention. I am not sure that men can even become friends without at least some verbal pushing and shoving.

Take the three tenors—Plácido Domingo, José Carreras, and Luciano Pavarotti—who were so successful that they filled whole stadiums. Their secret ingredient was a cheerful combination of rivalry and friendship. Their magnificent voices helped, of course. These three men, when they were younger, had been fierce competitors for the world's grand opera stages and so had every reason to dislike each other. When they began to sing together, they still sparred on stage about who was king of the high C's, but they also joked and slapped each other's backs like true friends. As Carreras said in an interview, "There was competition each time we went on stage. This is normal. At the same time, we had been really good friends. I can tell you that we had a lot of fun backstage!"[1]

This mix between contesting and getting along was so much a part of my growing up that it's second nature. The relations among my brothers were never as rough, however, as those described by the American author Tara Westover for her family:

> My brothers were like a pack of wolves. They tested each other constantly, with scuffles breaking out every time some young pup hit a growth spurt and dreamed of moving up. When I was young, these tussles usually ended with Mother screaming over some broken lamp or vase, but as I got older there were fewer things to break. Mother said we'd owned a TV once, when I was a baby, but Shawn had put Tyler's head through it.[2]

Like all boys, we were physical and had lots of shouting matches and scuffles even though I don't recall any life-threatening injuries. We played soccer, held Ping Pong competitions, skated on frozen canals, biked long distances together, and so on. Since moving up in the pecking order was not in the cards for me, my chief strategy was to break the ice. I am nonconfrontational and try to raise a laugh whenever I sense tension. I became a joker, at school and later in life. I may not look like one, because I carry the earnest face of my generation of Dutchmen, who routinely forget to smile in pictures. But my shtick has always been to find out what is funny about a situation.

This urge may strike at inappropriate moments, such as once when I burst out laughing in the middle of a serious academic workshop. Everyone looked at me with reproach in their eyes. I was reacting to a prominent anthropologist's assurance that our ancestors never mated with Neanderthals. His conviction came from the fact that those two hominids obviously didn't speak the same language even though they were physically extremely close. My mind, however, jumped to the international couples that I knew, including my wife and myself, who at their first meeting had few words available, only hands, lips, and a few other body parts. A decade later, the irrelevance of language in sexual affairs was confirmed when Neanderthal DNA was detected in the human genome.[3]

My attraction to the amusing side of arguments is one of the relics of being number four out of six. Another influence concerns food. I eat faster than most people and am not fond of leftovers. This is because in our home, we sat around the table with a pot of food in the middle. You had to ingest at a steady pace—otherwise all the food would be gone before you got your share. Leftovers were an unknown concept. A comparison with wolves might apply here because my centenarian aunt recently told me that while visiting our home, she was shocked by our ravenous eating habits. She lost count of the loaves of bread, liters of milk, and kilos of potatoes that were brought to the kitchen table and vanished in no time.

The special energy needs of boys are worth pointing out since a French feminist has claimed that the sole reason boys grow taller than girls is that they are favored at dinnertime. Nora Bouazzouni published a book cleverly entitled *Faiminisme* (a play on the French word *faim*, which means "hunger") in which she argues that humans are exceptional among mammals by having males larger than females. She ascribes this difference to parents depriving their daughters of food that they instead give to their sons. It's one of those biology-be-damned fantasies about gender. Bouazzouni not only got her mammalian biology wrong (males are larger in many species), she also underestimated the voracious appetite of boys. She should have visited my family when we were growing like beanstalks.[4]

In boys' peak year of growth, which is age sixteen (for girls it's twelve), they take in one and a half times as many calories as girls. These differences are driven by sex hormones, such as testosterone and estrogen, over which parents have no control. Whereas in prepubertal girls and boys, the ratio between body fat and muscles is similar, this changes drastically during adolescence. Boys gain lean body mass (bones and muscles), whereas girls add fat.[5] As a result, boys grow taller than girls. Naturally, different growth patterns require different nutrition. I'm sure my parents would have loved for us to eat a little less, but in the end, my mother was proud to be surrounded by sons who, like her husband, towered head and shoulders above her.

I have to think of her whenever people claim that we are a male-dominated species. In society at large, this may be true, but at home, my mom was boss despite her diminutive stature. We sometimes called her "the general" because she commanded a whole army to cut the bread, peel the potatoes, wash the dishes, go to the store, and so on. We followed a strict, heavily negotiated duty roster pinned to the wall. Her dominance gradually moved from physical to psychological, where it remained for the rest of her long life. For me, this transition occurred when I was about fifteen. I don't recall my father ever hitting us, but my mother would occasionally box an ear when she got angry. One day we were alone in the kitchen when she tried to slap my face, which was already above hers. I grabbed her arm and held it still in the air. We stood there both laughing because it was such a comical standoff, which left no doubt that the time of her being able to hit me had passed.

EVERY FAMILY HAS its gender composition, and for the author of a book on gender, there probably is no ideal one, but as a son from a family with a 7:1 sex ratio, I'm at a particular disadvantage. All things feminine remained mysterious to me for the longest time. I heard about menstruation or the growth of breasts, let alone sexual intercourse, only very indirectly and always veiled in euphemisms that were hard to figure out. The only thing my mother always said when she talked of girls or women is that we boys should respect them. She also didn't tolerate negative generalizations, whether they came out of my father's mouth or ours.

I normally don't dwell on my personal life, but a discussion of gender requires at least some background. I went to an all-boys elementary school, but even in high school, girls were scarce. In my class of twenty-five, there were only two. It was only when I went to college that I began to meet girls in larger numbers. My sexual development was late, as it was for most of my generation. In the beginning, my relations with young women were limited to studying together or discussing existential questions while listening to loud pop music (a bad combina-

My mother surrounded by her seven men. Being from a family with such a male-biased sex ratio probably drove my curiosity about gender issues.

tion, I'd say now), with an occasional dance party at which we pressed close, fumbled, and kissed. The first time a female friend visited my room for study, my landlady came up the stairs at least three times to knock on the door and ask if we wanted tea, something she never did when my male friends visited. I was seventeen at the time.

What impressed me most about girls was that they were so much nicer and sweeter than boys. Physically, of course, they could be incredibly soft and gentle in a way that was both new and delightful to me. But they also sympathized with me in a way that I had never experienced with my brothers or male friends. I had gained tons of the latter at the university. If a fellow student was disappointed (having failed an exam, undergone a breakup, or been kicked out of his room), we tried to cheer him up, punched his shoulder, came up with a solution, or distracted him with jokes. We'd raise a beer wishing him luck. We were supportive and helped if we could, but we didn't commiserate. We were not in the habit of offering a shoulder to cry on.

Women were different because if I had a setback, instead of trying to move me past it, make me forget it, or propose a way out, they would share my feelings. They listened, understood, offered calming contact, and showed concern. They might even get angry on my behalf, blaming that stupid professor for my shortcomings. This may sound like a stereotype, but it was what struck me most when I got to know women better. Their soothing reactions contrasted with what I was used to from male friends. Given my later interest in animal empathy, where similar sex differences are observable, this first impression has always stayed with me.

My studies gained in importance as my time at the university progressed. A few years into it, I had a chance to work with chimpanzees on the top floor of a tall building where two young males were kept in a separate room amid offices and classrooms. Such housing conditions would never be permitted today. Apart from a research project on memory, I also conducted my first gender experiment, but rather as a prank. I got the idea because the two apes, having no females of their species around, would show a prominent erection every time they saw a woman walk by but never for men. How did they detect human gender? A male fellow student and I tried to fool them by dressing up in skirts and wigs. We walked in, chatting in high voices and pointing at the chimps as if we were impromptu female visitors. They barely looked up, however. No erect penises, no confusion, except that they pulled at our skirts as if to say *What's the matter with you?*

How did they know? Smell was an unlikely cue because apes' senses are like ours: vision is dominant. Lots of animals easily distinguish human genders, though. Even species quite distant from us do so, such as cats and parrots. I know many a parrot who likes only women or only men and tries to bite the other sex. It's unknown where these preferences come from, but one general difference applies across the board: male movements tend to be brusque and resolute whereas those of females are more swinging and supple. This difference marks all sorts of species, including us. We don't even need to see a body to make this distinction. Scientists attached little lights to the arms, legs, and pelvis of people and filmed them walking. They found that just from watching a few moving white dots against a black background, we know which gender is walking by. This information is apparently sufficient. I bet that animals pick up on the same difference in movement.[6]

After my work with the chimps—to which I returned years later—I moved on to my favorite bird. The jackdaw is a gray-necked black corvid, a small member of the crow family. Abundant in European cities, jackdaws nest in church towers and chimneys. I love their happy metallic *kaw-kaw* calls while they fly around in pairs. I am such a romantic that I take pleasure in their lifelong pair-bonds even though science has uncovered that this arrangement is not as perfect as it seems. Offspring are not always fathered by the male of the pair, even though he dutifully defends the nest and feeds the chicks. Biologists contrast *social monogamy* with *genetic monogamy*. Because bird lives entail so much messing around, genetic monogamy is about as rare as it is in human society.[7]

Mated jackdaws call to each other in flight, when they land, and when they are about to take off. They always travel together except when they have eggs or young in the nest. The two of them step dapperly around in the grass while bobbing their gray heads, occasionally jumping up to catch a flying insect. They are rarely more than a few meters apart. We studied a whole noisy colony of these birds, which occupied nest boxes attached to a university building. The sexes have a clear task division while building nests. Both mates haul nesting material, the male the longer branches and the female the soft bedding, such as little twigs, feathers, and hair stolen from nearby horses and sheep. The female sometimes corrects her mate's efforts. If he keeps enthusiastically adding branches and the nest box gets too cramped, the female flies out carrying a big branch to drop it away from the nest.

As a university student, I joined a feminist organization, except that we didn't use this word. The keyword at that time was "emancipation." The organization was called Man Vrouw Maatschappij (MVM), which is Dutch for Man Woman Society. This national movement sought to improve the position of women while enlisting men as allies. It tried to realize its goals through political channels rather than the manifestations and protests that became popular later. I was recruited into MVM by the wife of a professor I knew.

Initially, I was entirely on board with its agenda. The idea was for women and men to work hand in hand to promote a new role division in society, one that would allow women more freedom and opportunities. Typical topics were reproductive rights, careers and jobs, income inequality, and political representation. These themes remain topical today. I am still convinced that progress requires male involvement— not because men are so brilliant or effective, but because the established order will not budge without sympathizers among those in power. This was true for the civil rights movement and will be true for women's liberation.

I quit MVM after a year, though, because the movement became increasingly hostile to men. Men were the villains and the source of all problems. In our discussion groups, the male minority would occasionally try to counter the rising animosity by pointing out that many men are hardworking providers for their families, or that every child needs a father and that men enjoy this role. These arguments were dismissed as irrelevant. Didn't we know that men rape? That they beat their wives? I was disappointed by the generalizations, especially after all the warnings against them with respect to women. They were all the more puzzling given that the women of MVM, who were mostly middle class, never complained about their husbands. Those men were apparently fine. It were the others who were trashed.

I simply refuse to turn against my gender. Books by a few male anthropologists have done so, such as *The Natural Superiority of Women* by Ashley Montagu, and *Women After All: Sex, Evolution, and the End of Male Supremacy* by Melvin Konner. The latter author treats maleness as a birth defect, calling it the "X-chromosome deficit." I have no taste for self-flagellation, however, and don't think we need to denigrate one gender in order to elevate the other. Most male members of MVM felt the same, and we left the organization in droves until all of us were gone. A few years later men weren't allowed as members anymore. That was when both of the women founders of the movement abandoned ship, too. Curiously, the organization kept its name even though the first *M* had become obsolete.[8]

After my brief flirtation with activism, I was lucky to meet a young feminist from the land of Simone de Beauvoir. At the time, however, I was hardly interested in the ideological side of our encounter. Catherine was twenty-one, and I twenty-two when we fell in love. The fact that we are still together demonstrates how great a match we were despite both of us being headstrong and domineering.

Perhaps our biggest difference is cultural. The Dutch pride themselves on being sober-minded and matter-of-fact, whereas the French are passionate and vocal about love, food, politics, family, and nearly everything else. The contrast in national temperament is a bit like comparing a movie by Ingmar Bergman with one by Federico Fellini. While I was getting used to Catherine's ardent spontaneity and strength of feelings, some of my Dutch friends felt intimidated and worried about my well-being. It never occurred to me, though, to attribute our differences to gender, such as the common generalization that women are more emotional than men. Since I view myself as driven by emotions and intuitions, I have trouble seeing this as gender-specific, let alone as a problem.

We have emotions for good evolutionary reasons. They guide an organism's behavior toward survival, hence their presence in all animals. Every animal needs fear, anger, disgust, attraction, and attachment.[9] Emotions are no luxury. Their relevance also does not vary much by gender. The emotions are quite rational in that they often know better what is good for us than our vaunted reasoning capacity.[10] In the West, however, we celebrate the latter and look down on the former. We view the emotions as too close to the body, which drags us down ("the flesh is weak"). The belief that men are more cerebral and less affected by emotions permeates popular culture, self-help books, and sitcoms. In an attempt to soften the blow, women may be said to have greater "emotional intelligence." This seems like a backhanded compliment, however, that still insists on a difference with men, who don't need all those feelings. It's no accident that the word *hysterical,* which denotes an unhealthy level of emotion, derives from the Greek word for uterus.

There is no scientific evidence, however, that the genders differ in the degree to which they follow emotions. One only needs to watch men during a crucial sports game to recognize their highly emotional nature. Even those stoic Dutchmen go bonkers as soon as they see an orange shirt run across the green grass! Gender differences mostly concern the triggers and strengths of specific emotions and the cultural *display rules* surrounding them, which tell us when it's appropriate to laugh, cry, smile, and so on.[11]

Display rules permit women to express more tender sentiments, such as sadness and empathy, and men more power-enhancing ones, such as anger. When a man raises his voice—as Supreme Court nominee Brett Kavanaugh did before the Senate Judiciary Committee, in 2018—his tantrum may be hailed as rightful indignation. Women, by contrast, often bite their tongues because they know that anger doesn't make them look good. In an actual experiment on this contrast, subjects on an imaginary jury were asked to come up with a verdict. The deliberations were done by text in a chatroom and sometimes got heated. If angry language came from a person with a male name, it amplified his point of view. But if the same words seemed to come from a woman, they undermined her credibility.[12]

The bias against emotionality is curious because it is by now well established that human thought, including that of men, is largely intuitive and subconscious. We can't even make decisions unless we have an emotional stake in them. As the Irish playwright George Bernard Shaw put it, "It is feeling that sets a man thinking, and not thought that sets him feeling." But even though everything begins with the emotions, the Western myth of rational man persists.[13]

Aᴏᴛᴇʀ ɪ ᴍᴇᴛ Catherine and her French family, and emigrated with her to the United States as a married couple, I was intimately familiar with three cultures. Each approached gender issues in its own fashion and moved at its own speed in relation to the job market, sexual morality, and education. Each culture was a mixed bag of progress.

Consider the French. In one of the foundational tracts of modern feminism, *The Second Sex,* which appeared in 1949, de Beauvoir noted that "one is not born, but rather becomes, a woman." This oft-quoted line has been interpreted to mean that womanhood transcends biological needs and functions. But it doesn't deny any of those needs and functions. They were taken seriously enough by the author's home country that it offered working women affordable childcare and generous maternity leaves. France was among the first nations with subsidized daycare centers (*crèches*), preschool programs, and home care for infants and toddlers. De Beauvoir herself cared enough about the specific needs of women to join the struggle for birth control and abortion rights.[14]

The Netherlands has always been marked by liberal sexual mores, even though conservative religious minorities remain. It was the first nation to legalize gay marriage. It also has among the world's lowest teenage pregnancy and abortion rates thanks to its sex education, which starts at the age of four.[15] Rather than scaring children and promoting abstinence, Dutch sex education seeks to foster mutual respect and to stress the pleasurable and loving side of sex.[16]

Despite their egalitarian gender ethos, however, the Dutch are not ahead in every way. With regard to the financial independence of women and their access to high-paying jobs, they are behind. I am always amazed, for example, by how few female professors I meet at Dutch universities. Two out of every three women with a job work only part time (the highest percentage in the industrialized world), one reason being the social pressure to take care of their family. The typical guilt trip is that one can't be both a good mother and work full time.[17]

In the 1980s, when we moved to the United States, we encountered an unusual mix of progress and conservatism. The country's sexual morality seemed stuck in the 1950s, yet in terms of education and jobs, women were more liberated. To enter the United States, I had to fill out a form declaring that I was neither a communist nor a homosexual— a requirement that was dropped only in 1990. This immediately signaled the conservative atmosphere we were to enter. For example, we learned of a custom called "proposing" that precedes marriage. American women wait, sometimes for years, for men to fall on one knee while holding up an expensive ring. Afterward, the lucky woman shows off her sparkling rock to the oohs and aahs of her friends. Marriage proposals were standard in Europe during the time of my grandparents but were aimed more at the parents of the future bride than at the bride herself. I realize that Americans consider it a perfectly fine ritual, a happy one even, but its blatant gender asymmetry floored us.

We also never got used to our adoptive country's prudishness and nipple obsession. Fear of the nipple has given rise to the uniquely American invention of the "lactation room," where women nurse or pump breast milk behind closed doors. Paid maternity leaves would make these rooms obsolete. So would tolerance of public breastfeeding, which is treated almost like a sexual act. Pictures of nipples are censured, bras are mandatory, and there was an actual "nipplegate" that lasted just half a second. Following Janet Jackson's breast exposure, in 2004, commentators lamented the nation's moral decline. The video of her "wardrobe malfunction," as it was called to avoid mentioning her body, was the most-watched of all time. It is said to have inspired the creation of YouTube.[18]

This fixation took us by surprise because in Europe breasts are no big deal. They are openly visible on primetime television, in mainstream magazines, in ads on the city bus, and live on the beach. Bras are mostly for support, not for concealment, and quite a few women go without them. If a baby gets hungry at a school meeting, at a party, or in the park, a breast will see the light of day to serve its intended function, even though outside the family, mothers usually first ask if everyone is okay with her doing so.

In Paris in the 1990s, the absence of nipple stigma caused a culture clash when the Disney company arrived with a strict dress code for its employees. Its insistence on "appropriate undergarments" led to street protests. With typical French hyperbole, the newspapers dubbed it an "attack on human dignity."[19]

Notwithstanding its sexual conservatism, the United States is well ahead of other Western nations with regard to the education of women, their participation in the workforce, and protections against sexual harassment. College education for women arrived earlier, and many women have made academic careers. Some academic disciplines have reached gender parity in their faculty, which means that search committees don't pay much attention to gender anymore. The rules concerning harassment have changed dramatically, too. They concern not merely unwanted sexual advances but also mutually consensual dating between people in the same organization, especially those with a power differential. The rules have changed so fast that a few prominent European politicians got caught unawares during visits to the United States. They were accused of the sort of lewd behavior that they could probably get away with in their home countries. With the #MeToo movement, protests against unwanted sex have only picked up steam, the impact of which is also felt in Europe.[20]

U.S. sexual morals are evolving in ways that I wouldn't have dared to predict a few decades ago. Cohabitation between unmarried couples is on the rise, birth out of wedlock is more common and more accepted, and same-sex marriage is legal nationwide. Tolerance of public breastfeeding, too, is increasing by the day. If a nursing mother gets shamed out of a restaurant, an angry mob of moms will show up the next day for a "nurse-in." The political momentum in favor of paid maternity (and paternity) leave will soon send lactation rooms the way of the dinosaurs.[21]

Ape breasts may reach B-cup size in nursing mothers, but they are deflated in the interval between offspring. Human breasts are unique in

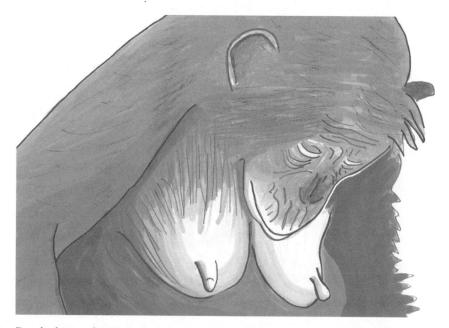

Bonobo breasts don't serve as sexual signals. Being swollen during periods of nursing and less hairy than the rest of the body, they can be quite conspicuous.

that they stay permanently swollen. We have sexualized these quintessentially mammalian organs, but this isn't typical of all human societies, and it has no equivalent in other animals. No dog ever gets excited by another dog's teats even though she has no less than eight of them. Breasts never turn the heads of male apes the way female behinds do.

Breasts are for nourishment, which is why young bonobos and chimpanzees get so attached to them. At the slightest disturbance or frustration (a lost fight with a peer, a bee sting), they run to Mom to suck on a nipple until they've calmed down. Apes typically nurse for four years, sometimes five, but the champion is the orangutan, which in the wild nurses for seven to eight years. Evidently, we are not the only slowly developing hominids. Wild apes have few resources available for their offspring other than the fruits in the forest, which the young start eating when they reach one year. The fruit supply isn't reliable, though, hence the need for an extended nursing period.[22]

When breasts don't work the way they should, we humans have

solutions. Wild primates don't have such options, but in captivity we can teach an ape to raise a baby on the bottle. I did so once with a chimpanzee named Kuif, to whom we gave a baby ape for adoption at the Burgers' Zoo. Kuif had lost some of her infants due to insufficient lactation. Every time she had gone into a depression marked by withdrawal, heart-breaking screams, and loss of appetite. With bars between us, I trained Kuif to handle a bottle and feed a baby chimp named Roosje, whom I kept on my side. The most challenging part was not to teach Kuif how to handle the bottle, which for a tool-using ape isn't that hard, but to make clear that the milk was not for her but for Roosje. Kuif was so extremely interested in the infant that she did everything I wanted and was a quick learner. After her transfer, Roosje clung permanently to Kuif, who raised her successfully. A few times a day, she'd come in from the outdoor island with her baby for a feeding session.

Kuif was eternally grateful to me. Every time I visited the zoo, sometimes after several years, she'd welcome me like a long-lost family member, groom me, and whimper if I made signs to leave. Later, the training also helped her raise her biological young.

Today few of the original chimpanzees survive in the Burgers' colony to greet me when I visit. Roosje is still there and has a daughter of her own. She doesn't know who I am, though, because she was a baby when I held her in my arms forty years ago. A picture of me doing so always draws guffaws, because not only do I look much younger, but I have long hair. My generation was massively protesting the authority of parents, universities, and government, and our hair and clothes signaled the revolt. In the evenings, I'd listen to bohemian-looking ideologues holding forth on the evil of hierarchies, while during the days, I'd observe power plays in the chimpanzee colony. This alternation induced a serious dilemma due to the contradictory messages.

In the end, I found behavior to be so much more convincing than words that I placed my trust in the chimps. I'm glad we can watch them without being distracted by them telling us about themselves. When it comes to power, their interest is plain. A particular male may have been alpha for years, but his position will inevitably be under fire from younger males. Actual physical confrontations are rare, and power struggles are mostly decided by coalitions in which two or three males band together. A challenger will approach with all his hair on end, throw objects at the alpha male to see how he reacts, or charge closely past him to see if he jumps aside. Any weakness or hesitation will be registered. An alpha male needs nerves of steel to withstand these provocations and develop counterstrategies, such as grooming buddies who support him. All these tensions play out over months, laying bare the enormous ambition to reach the top that is present in nearly every male of prime age.

And not just in the males. Mama, the colony's longtime alpha female, would underline her position vis-à-vis other females in no uncertain matter. She kept them in line to support her favorite male contender for the throne, acting as a party whip. If a female backed the "wrong" male during a status struggle, Mama might later in the day show up with her sidekick, Kuif, to deliver a severe beating. Mama didn't accept disloyalty.

I followed these dramas with utter fascination and began to read outside standard biologists' fare to understand what was going on. I got inspiration from Niccolò Machiavelli's *The Prince,* a book from half a millennium ago. The Florentine philosopher offered an insightful, unadorned account of the politics among the Borgias, the Medici, and the popes of his day. As a result, I also got a different perspective on the human behavior around me. Despite all their talk of equality, my fellow revolutionaries exhibited a distinct hierarchy, with a few driven young men at the top. Even though many women took part in the student movement, gender rarely came up in the calls for a new order. Women might wield power as the on-and-off girlfriends of male leaders, but they hardly did so on their own. This contradiction recalls the longstanding debate about egalitarian hunter-gatherers. To label those societies "egalitarian" requires one to disregard the pervasive status difference between men and women. One reviewer of the anthropological literature sarcastically spoke of "the belated discovery that foraging societies consisted of two sexes."[23]

True egalitarianism is indeed hard to find, and our student protest movement was a case in point. The front man had a habit of showing up late to mass meetings and striding into the auditorium followed by his acolytes. It was as if the king had arrived. The buzz in the room would die instantly. While we waited for him to take the podium and do his rabble-rousing, members of his inner circle would perform a warm-up act. They'd discuss less critical topics and practical matters, such as how to use the stencil machine. I witnessed several occasions when a young man in the audience stood up to point out inconsistencies in our position or criticize a particular decision. It was clear from the way his remarks got mocked and his ideological purity questioned that open debate was fine only so long as it didn't wobble the established order.

We collectively suffered an *egalitarian delusion*. We engaged in fiercely democratic rhetoric, yet our actual behavior told a different story.

I HAD TO THINK back to this delusion when I joined Emory University's psychology department. This was my third major transition: first from student to scientist, second from the Netherlands to the United States, and now from being surrounded by biologists to the world of psychology. Used to taking observable behavior as my starting point, I now had colleagues who presented human subjects with questionnaires and trusted their answers. I had entered an environment where the spoken word prevailed.

I learned a tremendous amount from my colleagues about human behavior. Most of them were excellent scientists, always critical of received wisdom, asking for data, and questioning common preconceptions. But because psychologists have the handicap of dealing with the species they belong to, they have trouble taking distance. They are right in the thick of what they study, which makes it hard not to judge behavior by cultural, moral, or political standards. This explains why psychology textbooks read almost like ideological tracts. Between the lines, we gather that racism is deplorable, sexism is wrong, aggression is to be eliminated, and hierarchies are archaic. For me, this was a shock, not because I necessarily believe the opposite, but because such opinions interfere with science. I may want to know how the races perceive each other or how the sexes interact, but whether their behavior is desirable is a separate issue. The task of science is not to judge behavior but to understand it.

Every time I received a psychology textbook from a publisher, I made a point of checking the index for entries on power and dominance. Most of the time, these terms were not even listed, as if they didn't apply to the social behavior of *Homo sapiens*. If they were included as topics about which students needed to learn, it was usually about the abuse of power or the drawbacks of hierarchical structures. Power was treated as a dirty word that deserves scorn rather than attention. This bias also explains the poor reputation of Machiavelli. Most scholars make a show of holding their noses while mentioning him. They would rather shoot the messenger than listen to him.

The egalitarian delusion of the social sciences is all the more astounding since we all work at a university, which is one huge power structure. It runs from the lowly undergraduates to the graduate students, the postdocs, the lecturers and professors of various ranks, all the way up to the "deanlets" (associate deans), deans, the provost, and the president. And within this structure, all of us are busy trying to expand our influence while limiting that of others. This activity is far from hidden even if the motives behind it are usually dressed up as something else, such as serving the needs of students or doing what is best for the university.

I have learned much from observing power plays among my colleagues: their divide-and-rule strategies, the formation of cliques, the silent nods of agreement when a rival is being criticized in a meeting, and even outright overthrows. At a critical meeting, one senior professor, who acted like the silverback of our department, was undermined by a coalition of junior faculty members whom he deemed his protégées. They must have plotted their coup because it came out of the blue. After the vote that marked his defeat, I never heard this professor's booming voice again. He roamed the hallways like a zombie, deflated. He retired within a year. I had seen it all before, only in another species.

The similarities are striking enough that my first book for the general public, *Chimpanzee Politics* (1982), drew the attention of U.S. Speaker of the House Newt Gingrich. After he put my book on the reading list for Congress members, the *alpha male* label began to gain currency in Washington, D.C.[24] Unfortunately, the term's meaning narrowed with time. It came to stand for leading men with an obnoxious personality. Alphas are bullies who never cease to let everyone feel who's boss. Current titles in the business book section are telling, such as: *Become the Alpha Male: How to Be an Alpha Male, Dominate in Both the Boardroom and Bedroom, and Live the Life of a Complete Badass.*[25] The popular image of the alpha male doesn't fit the way primatologists use the term, however. The alpha male is merely the top-ranking male regardless of how nicely or horribly he behaves. In the same way, every group also has an alpha female. There can be only one alpha of each sex. Most of the time they aren't bullies but rather leaders who keep the group together.[26]

Their unique position unexpectedly popped up in one of our behavioral experiments. Wanting to know if chimps care about the well-being of others, we tested them in pairs. One individual could choose food either for the two of them or only for him- or herself. Not only did the apes overwhelmingly prefer outcomes that had both of them munching, but the most generous individuals were the top-ranking ones of both sexes. Experiments with monkeys produced similar results. Why are alphas more prosocial than everybody else? It's a chicken-and-egg problem. Do these individuals get to the top by being helpful to others? Or does having a comfortable position make them more willing to share? Whatever the reason, this finding demonstrates why social dominance can't be reduced to being a bully. It's far more complex and includes generosity.[27]

Since the discovery of the pecking order in chickens, a century ago, we know how ubiquitous social ladders are in the animal kingdom. You put a dozen goslings together, or puppies, or monkeys, and battles for dominance are guaranteed. The same is true for human toddlers on the first day in daycare. It's such a primordial urge that we cannot wish it away. Yet we do. We discuss power as something that possibly appeals to others but certainly not to us. Three decades as a psychology professor have taught me that even serious scientists block out behavior that is right under their noses. Power remains a taboo topic, and we certainly don't like to hear how similar we are, in this regard, to other species.

We apply the same self-delusion to gender differences. We get so carried away by our hopes for the world that we forget what our actual behavior looks like. Some authors exaggerate the significance of gender along the lines that men are from Mars and women from Venus. Or that women are emotional and men are rational. But we also have those who, in reaction perhaps, minimize the differences to the point that they evaporate. Existing differences are presented as superficial and easy to get rid of. That neither extreme fits the evidence has become hard to appreciate among all the noise surrounding this issue.[28]

Perhaps we need to do what I typically do during political debates on television. I turn off the sound so that I can focus on the body language, which I trust more than the sound waves emanating from the candidates' mouths. In the same way, we should temporarily silence the voices in our heads that tell us how we'd like to see the genders behave and just watch how they actually do.

4

THE WRONG METAPHOR

Exaggerating Primate Patriarchy

Whatcould possibly go wrong?

What could possibly go wrong if you release one hundred monkeys into a large rockwork enclosure? Especially if they belong to a passionately harem-holding species and if, instead of releasing several females per male, you release an overwhelming majority of males with just a handful of females.

This experiment was carried out a century ago at Monkey Hill in Regent's Park Zoo in London. It didn't go well. The resulting mayhem and bloodbath became the basis of how the general public has viewed the intersexual relations of primates ever since. This was doubly unfortunate. Not only was the monkey species in question quite distant from us, but its behavior at the zoo was patently pathological. The hamadryas baboon—a large dog-faced monkey worshiped in ancient Egypt—has males twice the size of females equipped with long, sharp canine teeth. Moreover, the males grow a thick silver-white cape, whereas the females stay brown all over so that the males stand out even more.

Each male seeks to build a small polygynous family. At Monkey Hill, they fought fiercely over the few females, allowing their prospective mates no time to relax or even eat. They dragged their prizes around, killed some of them in the process, and copulated with their corpses. The zoo added more females, but that didn't stop the carnage. About two-thirds of all the baboons died, leaving a relatively calm masculine community after the battles subsided.[1]

Gender comparisons between ourselves and other primates thus started off on the wrong foot. It also didn't help that this foot belonged to an arrogant British lord who liked to throw his weight around and berate others. Solly Zuckerman, the zoo's anatomist, single-handedly "baboonized" the gender debate. He proposed that males are naturally superior and violent and that females have hardly any say at all. Females

The influence of primatology on gender debates got off on the wrong foot with extrapolations from hamadryas baboons. The possessive male in the background is about twice the size of the females around him. A silver-haired cape sets him further apart.

exist only for the males. In his 1932 book *The Social Life of Monkeys and Apes,* Zuckerman presented the events at Monkey Hill as emblematic of simian society and, by extension, ours.

Apparently unaware that herding and male control over females are atypical among primates, and oblivious to the fact that hamadryas baboons show an exceptional size difference between the sexes, Zuckerman freely adopted these baboons as avatars depicting the origin of human civilization, including our "compromise" of monogamy. Exaggerating the importance of sexual relations, he wrote, "The sexual bond is stronger than the social relationship, and an adult male, unlike a female, is not owned by any individual fellow."[2]

Few primatologists agreed, and by the time I started my studies, Zuckerman was largely forgotten. Yet his writing had a lasting impact on the general public. The assessments of this bellicose man, who in later years advised the British military on bombing raids, have seeped into popular culture, and we have been unable to keep them out. His account was just too compelling. Or perhaps it was too much in line with how people *wanted* to see, or were accustomed to seeing, themselves. We say that nature acts as a mirror, yet we rarely use it to see anything new. After the horrors of World War II, people were prone to believe in their own depravity. Monkey Hill reinforced their dismal self-assessment and became grist for the mill of scores of authors who regarded humans as evil "killer apes" engaged in a Hobbesian struggle of all against all.

The Austrian ethologist Konrad Lorenz told us that we lack control over our aggressive instincts. Not long afterward the British biologist Richard Dawkins stated that our chief purpose on earth is to obey our "selfish genes." Even our positive traits had to be phrased as if they were suspect. So if animals and humans love their families, biologists preferred to call it "nepotism." The baboon drama at the zoo was compared with the mutiny on the *Bounty,* an eighteenth-century maritime rebellion in which thirty men on an island ended up killing each other. It was echoed in William Golding's 1954 novel *Lord of the Flies,* in which

British schoolboys descend into a near-cannibalistic orgy of violence. These and other books gleefully presented our species as mean, cruel, and morally bankrupt. That's just the way we are, the authors shrugged, and anyone who tried to offer a more uplifting picture risked being ridiculed as romantic, naïve, or ill-informed. Anthropologists highlighting peaceful coexistence between tribes, for example, were quickly dismissed as "peaceniks" and "Pollyannas." Since Monkey Hill had laid bare the beast inside all of us, we'd better fall in line behind the ideas it had spawned.

It's remarkable how influential primate comparisons can be. Not satisfied with analyses of human behavior per se, we like to put them in a broader context that includes the sorts of animals our ancestors must have resembled. We don't even stop there, though, but go on to revel in allegories that strip away the role of civilization and connect us with apes at an emotional, even erotic level. Examples are *King Kong, Tarzan, Planet of the Apes*, Peter Høeg's *The Woman and the Ape*, and myriad other fantasies. We're unable to look away from the parallels. This is why Monkey Hill resonated so widely outside primatology despite the current evaluation that it was a case of gross mismanagement and cocky overinterpretation.

Zuckerman himself never shied away from academic fights. He trashed any colleagues who dared to argue that primates do *not* customarily kill each other, or that males and females typically get along. He also criticized those who claimed that primates possess remarkable intelligence and social skills. He saw himself as the only real scientist, the one who didn't sugarcoat human nature. All others were "anthropomorphic"— the curse word of choice in relation to animal behavior.

Nevertheless, Zuckerman was unable to stop a new generation of primatologists from coming up. In 1962, at the London Zoological Society, a twenty-something Englishwoman dared to question *Man the Tool-Maker,* a widely acclaimed book by the anthropologist Kenneth Oakley that had given us the definitive trait that sets humanity apart: not the use of tools, but our ability to manufacture them.[3] Jane Goodall, however, was a keen observer who had seen wild chimpanzees strip leaves and side branches off twigs to make them suitable for termite fishing.

Her lecture was well-received, except by Zuckerman, the secretary of the society, who gradually turned purple with indignation. My Dutch professor, Jan van Hooff, was present and recalls him throwing a fit, demanding to know from the organizers, "Who invited this unknown, ridiculous *girl* to a scientific meeting?"[4] Later, in a self-congratulatory piece in *The New York Review of Books* under the humble name Lord Zuckerman, he laid into those "attractive young women" who were taking over the field. He accused them of using anecdotes and "empty words" to describe the sort of well-ordered primate societies that the lord himself had never encountered.[5]

He didn't live to see Goodall being named a dame.

THIS HISTORY EXPOSES, in a nutshell, multiple tensions within our field: between studies in captivity and in the wild, between the male establishment and the first female primatologists, and between pessimistic and optimistic views of human nature. Before I consider the gender implications, let me briefly elaborate on the general mood change in biology and Western society over the last few decades. We have moved from total bleakness to a more optimistic view of human nature.

My biggest problem during the postwar period was the doom and gloom of its most celebrated thinkers. I didn't share their negativity about the human condition. I had studied how primates resolve conflicts, sympathize with each other, and seek cooperation. Violence is not their default condition. Most of the time, they live in harmony. The same applies to our own species. I was shocked, therefore, in 1976 when Dawkins asserted in *The Selfish Gene,* "Be warned that if you wish, as I do, to build a society in which individuals cooperate generously and unselfishly towards a common good, you can expect little help from biological nature."[6]

I'd argue quite the contrary! Without our long evolution as intensely social beings, we'd be unlikely to care for our fellow humans. We have been programmed to pay attention to each other and offer help when needed. What else would be the point of living in groups? Many animals do, and they do so only because group life, which includes giving and receiving assistance, yields tremendous advantages over a solitary life.

One time Dawkins and I politely disagreed in person. On a cold November morning, I took him and a cameraman up a tower at the Yerkes Field Station. It overlooked the chimps that I knew so well. I pointed out Peony, an old female. Her arthritis was so acute that we had seen younger females hurry to fetch water for her. Instead of letting Peony slowly trek to the water faucet, they'd run ahead of her to suck up a mouthful and return to spit it into her mouth, which she opened wide. They also sometimes placed their hands on her ample behind to push her up into the climbing frame so that she could join a cluster of grooming friends. Peony received this aid from individuals unrelated to her, who surely couldn't expect any favors in return because she was not in a condition to deliver them.

How to explain such behavior? And how to explain all the acts of kindness that we ourselves engage in every day, sometimes with complete strangers? Dawkins tried to salvage his theory by blaming genes, saying that they must be "misfiring." Genes, however, are little strings of DNA devoid of intentions. They do what they do without any goals in mind, which means that they can't be selfish or unselfish. They also can't accidentally miss any goals.

During the 1970s and '80s, the focus on the dark side got so bad that I compared my life to that of a toilet frog.[7] I had met a big one in Australia, living in a toilet bowl, holding on with its suction cup toes during the occasional tsunamis that humans produced. This frog didn't seem to mind the body waste swirling down the bowl, but I did! Every time a book came out on the human condition, whether written by a biologist, an anthropologist, or a science journalist, I had to hang on desperately. Most of them advocated a cynical view totally anathema to the way I viewed our species.

My only solace during those years came from reading Mary Midgley. Like David Hume before her, Midgley was distinctly animal-friendly and always insisted that humans *are* animals. We are ultra-social animals with solid communal values. Unimpressed by all this talk about the absence of charity, she took on Dawkins directly.[8]

I began to realize that the lack of trust in human nature came almost exclusively from male colleagues. It was not typical of any female scholars that I knew. The literature depicting humans as greedy individualists was written by men for men. Its ultimate inspiration came from man-made religions according to which we arrive in this world as sinners with a big black blot on our soul. Being good was a thin veneer that covered an utterly selfish agenda. I dubbed it Veneer Theory.[9]

Around the turn of the century, I was happy to see a stream of fresh data bury these ideas. Anthropologists demonstrated a sense of fairness in people across the globe. Behavioral economists found humans to be naturally inclined to trust each other. Children and primates exhibited spontaneous altruism without enticements. And neuroscientists found that our brain is wired to feel the pain of others. My early work on empathy in primates was followed by research on dogs, elephants, birds, and even rodents, such as experiments in which one rat could free a trapped companion.[10] We now realize that the preponderance of open competition in the natural world— the so-called struggle for life—had been grossly overstated.

Even fictional accounts, such as *Lord of the Flies,* came under fire. Whereas violence among people marooned on an island does occur, especially in combination with starvation, it is by no means the rule. Our species excels at conflict resolution. Studies by psychologists suggest that children, instead of needing supervision, have no trouble settling their disputes if the adults leave the room.[11]

They do so even under the circumstances imagined by Golding. The Dutch historian Rutger Bregman found a blog on the Internet that read, "Six boys set out from Tonga on a fishing trip. Caught in a huge storm, the boys were shipwrecked on a deserted island. What do they do, this little tribe? They made a pact never to quarrel." Bregman became curious about this incident and traveled to Brisbane, Australia, to meet the survivors, now in their sixties. As boys between thirteen and sixteen years old, they had spent more than a year stranded on a small rocky island. They managed to make fire and feed themselves from a vegetable garden while avoiding fights. They'd cool their tempers if tensions arose. Their story was one of trust, loyalty, and friendship that lasted for the rest of their lives. The message was quite the opposite of the one Golding had tried to impress upon us.[12]

Why do so many people still believe Golding's ghastly story? Why has his book become a classic at middle schools as if it yields significant insights into human nature? And why does Zuckerman's account of the slaughter at Monkey Hill still loom behind popular depictions of the "natural order" despite having been thoroughly debunked? Perhaps it's our fascination with bad news, or as the American novelist Toni Morrison put it, "Evil has a blockbuster audience; Goodness lurks backstage. Evil has vivid speech; Goodness bites its tongue."[13]

We have fallen for Zuckerman's false primate narrative of wretchedness, which divided the sexes into the rulers versus the ruled. Never mind that the rulers ended up empty-handed. All this served as a metaphor of human society, promoted by an abrasive man who knew how to stall the flow of new information. Fifty years on, Goodall remained traumatized, such as during an interview around her eightieth birthday: "At the mention of Zuckerman, Goodall's features sharpen slightly, and the pace of her speech quickens. She dismisses his monkey work as "rubbish." It is the only bad word she has to say about anyone."[14]

THE SCIENTIST WHO ultimately buried Zuckerman's storyline was the influential Hans Kummer, who worked all his life on the same monkeys, hamadryas baboons. He did so first at the Zurich Zoo and later in their native habitat, in Ethiopia. He was my hero when I was younger because he was rigorous, creative, and open to new interpretations. I read every article that he wrote and emulated him.

I met him for the first time in person as a budding student of primate behavior. During a conference in Cambridge, I was allowed to sit with a few big-shot professors at a dinner table. The dinner was held in one of those spacious Gothic-style dining halls of the old university. While we were introducing ourselves, and I was silently congratulating myself on my luck, something odd happened. Through the loudspeaker came the call for certain people, mentioned by name, to join the "high table." The whole concept of a special high table was alien to us, continental Europeans. It sounded offensive because it introduced a class division that no one had asked for. In the old days, the high table had chairs while all the other tables had benches, but I don't recall if this was still true. Kummer was among the invitees to this table. He laughed it off and said that he liked our company better. We had a great evening. His spontaneous gesture has always endeared him to me.

Kummer was methodical in his data collection, yet he was ready for surprises. He would tell us that he was suspicious of results that agreed too much with his theories. What could be more exciting than finding something that changes your mind? He was a bearded patriarchal figure, which was appropriate given the species he studied. At the start of his book *In Quest of the Sacred Baboon*, he warned against making too much of this animal's behavior:

> *Although the ancient Egyptians saw the hamadryas baboon as a sacred figure, it is no saint. Its social life is not the idyll we fondly hope to find among animals. It lives in a patriarchal community, in which the male has evolved both of the fundamental aspects of fighting: a sharp canine tooth and a network of alliances. . . . When I started my research, I wasn't looking for a patriarchal society, nor was I aware that this was one, and this book should by no means be taken as subliminal propaganda for male superiority. What animals do is no argument for what humans ought to do.*[15]

This reflection on male superiority showed quite a bit more nuance than Zuckerman's hype. Kummer was well aware that his baboons were a "feminist nightmare," as he once called them in a lecture. He wisely replaced the old "harem" terminology with one-male units (OMUs). His field studies showed how males try to avoid violence. They collect females, which they defend against each other, but they have a range of subtle signals to forestall clashes. They have great respect for each other's OMUs. Once a male and female have established a tie, other males rarely infringe upon it.

Apart from doing field observations, Kummer captured wild baboons to test them, after which he'd set them free again. This way he found out, for example, that if a female entered a cage with two males, the males would fight over her. But if the female was put with only one male, while the other could watch from an adjacent pen, the outcome was strikingly different. The female needed only to spend a short while with one male for the other to respect their bond upon introduction into their cage. Even a big, totally dominant male would be inhibited from fighting. Instead, he'd sit at a distance from the pair, fiddling with a pebble on the ground. Or he might attentively scan the landscape outside his enclosure, shifting his head as if he had spotted something incredibly interesting. Kummer, however, was never able to detect what such males had seen.

The weaponry of these baboons is so nasty that they are reluctant to resort to it. Kummer reported that if you threw a peanut in front of a single male baboon walking by, he would invariably pick it up and eat it. If you did the same to two males walking side by side, they would appear not to notice the peanut. Both of them would walk straight past it as if it didn't exist. A peanut was not worth a fight. Kummer also observed that males wouldn't even try to assert dominance if their respective families entered a fruit tree too small for all of them. Both males would exit the tree in a hurry with their families in tow, leaving the fruit unpicked.

This deep aversion to conflict makes clear what had gone wrong at Monkey Hill. By throwing individuals of both sexes together without any preexisting bonds or established order among the males, the finely tuned mechanisms that generally keep them from fighting had broken down.

Kummer found that male behavior is not the sole factor underlying OMUs. True, a male will punish a female who strays too far, delivering a neck bite, whereupon she will stay near him to avoid further trouble. Yet females are not mere property. Kummer's team discovered this by including female preferences in the above experiments. They presented each female with two males in separate cages to see which one she'd like the best. They measured how much time she spent next to each male. Then they paired her with one of the males. The more she had preferred him during the pre-tests, the more reluctant other males were to challenge their bond. Only if she was paired with a male who was near the bottom of her preference scale would other males try to steal her from him. Kummer spoke of male "consideration" of what the female wants, which he saw as "a first evolutionary step on the road to a more egalitarian society."[16]

It strikes me as an extremely modest step, though. And it remains unclear if the males are really that considerate. Maybe they're just making sure not to fight for a prize they can't keep. The males must be picking up on a female's preference: primates are excellent at reading their own species's body language. They may calculate that if a female likes another male better, she's bound to run off at the first occasion that presents itself. From other studies, we know that males are unable to herd females who refuse to be herded.

The deeper problem is that we're looking for human gender parallels in baboons, which are monkeys, whereas we are apes. You may think of yourself as something else, but genetically we are right in the middle of the small hominid family. We are not even a side branch. The hominid family is defined by the absence of a tail (whereas monkeys have tails), flat chests, long arms, large bodies, and exceptional intelligence. Apart from humans, this family includes chimpanzees, bonobos, gorillas, and orangutans. No one has ever given a good biological reason why humans shouldn't be called apes—or bipedal apes, if you wish. Some have even suggested that our genus should be merged with that of our closest relatives, chimpanzees and bonobos. For historical and ego reasons, however, we cherish our separate genus, *Homo*. But given the DNA similarities with other apes, it might be more appropriate, in the words of American geographer Jared Diamond, to classify ourselves as "the third chimpanzee."[17]

DESPITE THEIR DISTANCE from us, baboon studies have revealed the impact of female (and feminist) primatologists. Baboons are among the easiest primates to watch, which is why they were the first to attract widespread attention in the wild. They became the subject of hundreds

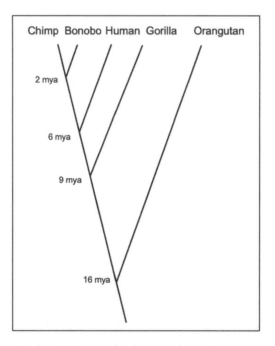

Based on DNA, our family tree indicates how many millions of years ago (mya) the five extant hominids (humans and the great apes) diverged. Since bonobos and chimpanzees split from each other long after we descended from their ancestor, they are equally close to us. The gorilla is a bit more distant, and the orangutan even more so. Hominids split off from the monkeys—primates with tails—around 30 mya.

of field reports, making them a test case of how a scientist's gender influences their approach.

I once followed a troop of baboons on foot in Kenya. It was a cakewalk compared to tracking primates in the forest. With arboreal primates, you are peering upward all the time, trying to see them through dense foliage. You will see only bits and pieces of their social life because they vanish as soon as there is a confrontation or dangerous situation. Only scientists who invest years in habituating forest primates to their presence manage to see more. Habituation takes patience, and in the early days, few fieldworkers took the time.

But baboons travel across the open savanna, always alert because of the dangers lurking in the tall grass. They are not particularly shy around humans with binoculars and a clipboard. They keep doing what they are doing, which is mostly foraging for herbs, fruits, seeds, roots, and the occasional antelope calf. They like meat but eat mostly plants. Almost the entire troop remains in sight of one another, even during a commotion, such as a fight. On the savanna, you can easily watch such events unfold, which is quite different from the situation in the forest.

Apart from the ease and convenience of watching baboons on the plains, primatologists had a second reason for focusing on them. Our forebears left the forest to enter the savanna, and baboons were an ideal model species because they followed the same path. They adapted to the same habitat. As primatologists repeated this ecological argument over and over, it almost seemed that there was no point in studying any other primates. This argument applied not only to the hamadryas baboons of Kummer and Zuckerman but also to closely related baboon species in which females are not "owned" by any males. In these baboons—divided into olive, chacma, and yellow baboons—females are autonomous. They form a cohesive, kin-based society that operates independently of the males. Their sons leave at puberty to join other troops, whereas adult males living in the troop have come from the outside.

During the masculine heyday of primatology, the emphasis was entirely on the pugnacious males. Described in almost militaristic terms, they were thought to serve a governmental function. The male hierarchy was the backbone of society that regulated every aspect of social life, including guaranteeing the safety of mothers and offspring.[18] Female baboons travel with their little ones jockey-riding on their backs, where they use Mom's tail as a backrest. Scientists described the progression of a baboon troop as resembling a battle formation: a large number of females and juveniles fearfully huddle in the middle, surrounded by males with formidable canines, ready to repel outside danger.

The first women primatologists to arrive, however, didn't see things the same way. For them, female baboons were the core of society. Female kinship networks are stable over time, reinforced by a great deal of grooming and soft grunting at each other's babies.

Among the early female baboonologists was the iconoclastic Thelma Rowell. I vividly remember this British primatologist because of the contrarian gleam in her eye. Her presence at conferences almost guaranteed pandemonium because while the men talked up the role of contest and status, Rowell would simply state that she'd never seen much evidence for it. She acted as an intellectual disrupter. By questioning the whole concept of social dominance, she instigated a lively debate that went on for years in the literature. Could it be, she'd ask, that by providing primates with a concentrated food source (a common field technique at the time), we were forcing them into a hierarchical pattern?

Rowell herself studied forest baboons in Uganda without setting out any food to draw them close. Her monkeys were relatively peaceful, and the males never warded off outside danger: "Baboon males are often described as defending their troop, but this I never saw and find difficult to imagine, since Ishasa baboons always reacted to any potential danger by flight . . . the whole troop flees from any major threat, the males with their long legs at the front, with the females carrying the heaviest infants coming last."[19]

Observations of how baboons react to predators are so scarce that one wonders how primatologists ever established the male's defensive role. Could it have been a product of human imagination? Only one study collected a fair number of predator episodes. The American anthropologist Curt Busse spent two thousand hours following chacma baboons in Botswana and camped close to their roosts at night. He discovered that leopards attack (and kill) baboons only in the dark. They never do so in the daytime. It is unlikely that male baboons can defend their troop against a nightly attack by such a daunting stalker even if during daytime they harass leopards.[20]

Busse also saw many encounters with lions in the daylight or at dusk. Lions are too large to be intimidated by baboons. Every baboon, large and small, responded to these cats by fleeing up a tree and giving loud alarm calls. Sometimes after a lion ambush was over, adult males would vigorously shake branches and bark at their enemy, but this was mostly for show. These observations, like those of Rowell, failed to support the heroic protector role, which by this time had nonetheless become a staple of anthropology textbooks.

Rowell had her own challenges as a female primatologist. When in 1961 she submitted an article under the name of T. E. Rowell to the journal of the Zoological Society of London, the society invited her to give a prestigious lecture to its fellows. The story goes that the fellows discovered what gender the article's author was only when Rowell walked into the room. An awkward situation ensued. The fellows had planned a dinner afterward, but felt they couldn't share a table with a woman. Unbelievably, they asked Rowell to eat behind a curtain. She declined.[21]

Rowell, like Goodall, was part of the first wave of female primatologists, soon to be followed by a much larger second wave. In 1985, two decades after the above controversies, the American anthropologist Barbara Smuts wrote my favorite baboon book, *Sex and Friendship in Baboons*.[22] The term *friendship,* however, raised eyebrows due to the cynical view of nature in those days. No one ever objected to allusions of animals having "enemies" or "rivals," but could they truly have friends? It would suggest that animals might like each other and be loyal—which was precisely what Smuts had documented in her baboons. Nowadays we know that having kin and friends is no luxury. It lowers the mortality risk of social animals, as it does in us.[23]

Instead of males imposing their will, as among hamadryas baboons, the intersexual friendships in savanna baboons are entirely voluntary and based on mutual attraction. The relations might be sexual, but more often they are platonic. Baboons start by "flirting" (surreptitiously stealing glances at each other) for a couple of days, then one makes a "come hither" face to the other, with lifted eyebrows and friendly lip-smacking, and they end by spending much time together. They travel, forage side by side, and huddle at night to stay warm. There is no coercion, only affection and trust. Every female in the troop has at least one male friend, usually two, and since there are fewer males than females, males often have multiple friends. Long-term resident males may have five or six female friends. For a female, being much smaller than a male, the great advantage is to have a powerful guardian. Her male friends defend both her and her infants against other females and especially against males.

I saw this for myself in the same Eburru Cliffs troop, in Kenya, that Smuts had studied. It's always easier to see a pattern once it has been reported—I was there a decade later. A young adult male, whom scientists had named Wellington, had immigrated from another troop just a few days earlier. He made all the other baboons extremely nervous with his cockiness and long sharp canine teeth that he regularly flashed through yawns and threats. They might chase him up a tree, but Wellington always returned to the ground. Older males have worn down or broken teeth, which is why they often band together to keep these young upstarts under control. When Wellington menacingly approached a female, she screamed and fled straight to a resident male, clinging tightly to him with both hands. Her friend stared down Wellington, who circled the two of them on high limbs, looking tall. He didn't dare touch the female, though.

Females trust their male friends around offspring. A mother may leave her infant in the care of her friend while she goes off to forage hundreds of meters away. A male babysitter offers far better protection than a sister or a sibling. Recently, genetic studies have shown that nearly half of the male friends have fathered offspring with the females they associate with.[24]

What a different impression of baboons we were getting! Female primatologists have taught us that relations among the females and the choices they make are an essential factor, equal to the effect of the male hierarchy. Females have a big say over whom they mate with, and they may even determine which males are welcome in the troop. It is not that male primatologists were wrong to stress male competition, which is plain to see in baboons, but that theirs was only half the story. The influx of women into primatology thus brought gender balance not only to our community but also to accounts of primate society.

W E S H O U L D N ' T B E surprised that the interests of scientists are gender-bound. Our approach is affected by every part of our background, including education, gender, academic discipline, and culture. More-over, a biologist naturally looks at animal behavior differently than a psychologist or an anthropologist. As for culture, my own fascination with conflict resolution relates no doubt to me coming from a small crowded country. In the Netherlands, consensus and tolerance are often valued over individual success. All of us bring a different perspective to the table.

But to argue from there that truth is elusive, and that reality is up for grabs, is profoundly misleading. It is a dangerous suggestion regularly encountered in the stream of books that romanticize women primatologists. This "beauty and the beast" genre celebrates Western women in the jungle as braver than men, nicer to animals, and communing with nature on a level that men can only dream of. It started in 1989 with Donna Haraway's *Primate Visions,* a postmodern analysis of primatology that remains a classic in the humanities. Not that there is anything wrong with honoring women, but if the implication is that there is no objective reality, it becomes questionable. Haraway's book suggested that the only reality is the one we are willing to see, and that women see a different (and better) one than men.

Her book set off alarm bells among primatologists, mostly among the men but also the women. Who likes to be sexualized along the lines of "White women mediate between 'man' and 'animal' in power-charged historical fields"? Or "The woman scientist of *National Geographic* is married to the camera's eye, or she is a virgin wise woman, married only to nature in the touch of a male ape"?[25] I don't even know how to read such fuzzy sentences dripping with insinuation. What I do know is that if I were a woman primatologist, I'd hate to be depicted as lusting after male apes. Instead of lifting up women, which was undoubtedly Haraway's intention, she undermined their authority by emphasizing their gender at the expense of their science. During one of the many animated debates on this topic, I heard one primatologist exclaim, "I don't want to be known as a female scientist! I want to be known as a scientist. Period!"

At least Haraway's book gave us a most entertaining takedown when the American anthropologist Matt Cartmill sank his teeth into it:

This is a book that contradicts itself a hundred times; but that is not a criticism of it, because its author thinks contradictions are a sign of intellectual ferment and vitality. This is a book that systematically distorts and selects historical evidence; but that is not a criticism, because its author thinks that all interpretations are biased. This is a book full of vaporous, French-intellectual prose; but that is not a criticism, because the author likes that sort of prose and has taken lessons in how to write it. This is a book that clatters around in a dark closet of irrelevancies for 450 pages before it bumps accidentally into its index and stops; but that is not a criticism, either, because its author finds it gratifying and refreshing to bang unrelated facts together as a rebuke to stuffy minds.[26]

The problem with Haraway's thesis is that scientists aren't looking for a pleasing "narrative" any more than that they want to feel at one with nature. If the latter happens, it's a bonus. Our primary goal is knowledge and explanations that will stand up to scrutiny. Postmodern scholars may believe that everyone has their own personal truth, but scientists believe in a knowable and verifiable shared reality. Except for Schrödinger's cat, there can be only one truth. It is our task to uncover it even if it doesn't meet our expectations, which is why the hottest term in science is *discovery*.

If the scientific enterprise were just a matter of us confirming our biases, we surely wouldn't need to work so hard. All we'd need to do is watch our primates for a couple of weeks and return with the story we wish to tell. We would have no need to spend years and years sweating in the field, living under primitive conditions, risking malaria, snakebites, big cats, and so on. We would also have no need to return with bags full of smelly fecal samples to be analyzed by a laboratory at home. Similarly, experimental scientists wouldn't need to come up with clever tests and the right controls to prove a specific point about the mental capacities of their subjects. Why conduct experiments at all if we already know the answer?

In the last few years, animals have demonstrated spectacular cognitive feats, few of which were known or even suspected twenty-five years ago. Remember, Kummer said that the most exciting results are those that take us by surprise. As soon as we find something new, we enter this delicate zone between what we believe exists and what we can say about it with confidence. All of us—men and women—are subject to the same rules of evidence. Haraway's allusion to primatologists "crafting" data is a shocking suggestion. Her choice of words comes close to fabrication! This is emphatically not what we do. Data are collected, not crafted.

Tellingly, the most cited paper in our field is one by the American ecologist and primatologist Jeanne Altmann in which she offers standardized methods for the observation of behavior. We first learn to identify and name every individual, which for a troop of a hundred baboons may take a few months. Then we follow them day after day to document their behavior under a wide range of circumstances. In the old days, we kept records with pencil and paper but nowadays we use digital devices.

Everything is codified, counted, tabulated, and graphed so that others can judge for themselves if our conclusions are on firm ground. The typical journal article is unreadable unless you know math and statistics.[27]

The late American primatologist Alison Jolly, a beloved senior figure and peacemaker in our community, tried hard to find a silver lining in Haraway's account. In the end, she was as exasperated as everyone else by the implication that wild primates are like a blank screen waiting for the projection of our prejudices. In her book *Lucy's Legacy*, Jolly tells of a meeting in Brazil on this issue among female primatologists. All the women agreed that science moves ahead by finding evidence that no one had expected—for example, that apes make tools, that male and female baboons can be friends, or that bonobo females dominate males—followed by battles with skeptics, because novel claims invariably invite disbelief:

> *Each of the field scientists claimed that our best-known discoveries were forced on us by the monkeys and apes themselves. In fact, many of us at first fought what we were seeing, because it contradicted our previous mindset. Of course, some mix of gender and funding, family and national background, got us into the field and prepared our minds for change, but we still saw something new. And of course when we did, that was what became famous, for it surprised others, too.*[28]

One day someone will write the history of our discipline without paying undue attention to the observer's gender. Primatology is now one of the few fields in science that is truly equal opportunity.[29] I do believe that the groundswell of women has expanded our horizon, but it has not fundamentally altered the way we conduct science. We still follow the same ground rules for acceptable evidence and insist on verifiable data and statistical analyses.

The primate order includes more than two hundred species, and it is unfortunate that early attention focused so much on the baboon. These monkeys are not the most suitable for exploring gender parallels with our species. The longtime baboon watcher Shirley Strum recognizes the limitations. Whereas her first popular book on baboons still carried the title *Almost Human,* Strum now emphasizes the many ways in which these monkeys differ from us.[30]

While baboons inspired the myth that patriarchy is natural and that macho males make up the core of society, we now know that this isn't true even for baboons, let alone for most of the other primates to which this idea was generalized. Fortunately, science has left these distorted notions behind, along with the widespread opinion that human nature is inherently nasty, brutal, and selfish. Over the years, we have undergone a radical shift in the way we frame the evolution of sociality. We now emphasize cooperation at least as much as competition.

One of the early advocates of this shift, Mary Midgley, was ahead of the curve, while I was caught in the middle. Midgley and I had a good chuckle at the many battles behind us when I recently visited her in a retirement home in Newcastle, UK. Although she was ninety-eight years old, she insisted on making me a cup of tea, while I could only pay homage to her life's work.

She died a year later.

5

BONOBO SISTERHOOD

The Forgotten Ape Revisited

Walking along a sand path around the enormous sanctuary, we are accompanied behind the fence by a male bonobo. His hair is on end, and he pulls a branch behind him while running past us. Then he comes back to run again. And again. Bonobos show off by dragging objects along the ground, which adds noise to their display. This male is doing so because my guide at Lola ya Bonobo (Paradise of the Bonobos, in Lingala) is its founder, the Belgian conservationist Claudine André.[1] She knows him; he knows her. Most bonobos here were in Claudine's arms when they were little. Besides, there is me. I am both a male and a stranger, which makes me a rival.

When we approach the next enclosure, a male from a different group takes up the branch-dragging. Given how lush and spacious these forested enclosures are (the largest one is almost forty acres), no bonobo has any reason to hang out with us. They have plenty of space. That males patrol the periphery signals their territoriality. When bonobo groups meet in the wild, males chase neighboring males. But since these encounters are typically initiated by females, who are eager to mingle and groom with their neighbors, male competition never escalates to the level of violence seen in chimpanzees. Chimps kill their enemies, whereas male bonobos barely make a scratch.[2]

The male's branch display comes to an abrupt halt when he spots a female in the next-door enclosure. She's walking around with a prominent genital swelling on her behind that is typical of her species. Usually a female in this condition will move nonchalantly, as if she doesn't notice all the male eyes glued to her derrière. However, this female briefly stands still behind the fence to shake her pink balloon, making it jiggle like a pudding. While doing so, she stares directly into the male's eyes as if asking, *What do you think of this?*

No wonder the bonobos at Lola have found ways of braving the electric fences to move between enclosures. One male swaps groups so often that no one knows anymore to which one he belongs.

Bᴏɴᴏʙᴏs ɪɴᴄʀᴇᴀsɪɴɢʟʏ sʜᴏᴡ ᴜᴘ in feminist discourse as humanity's last hope. Their existence is taken as proof that male dominance is not hardwired in us. I am okay with this conclusion as long as we keep in mind that we have an equally close relative, the chimpanzee, which is quite different. Each ape has its unique specializations, which makes it hard to extrapolate directly from them to us. Best is a triangular comparison involving our next of kin and ourselves to see what we have in common and where each one of us diverges.

Bonobo behavior violates popular notions about our heritage. At the very least, it hints at greater flexibility than was once widely assumed. Unfortunately, few people know what bonobos are and at best refer to them as "bonobo monkeys." They are not monkeys, however, but tailless hominids like us. This gives them a special status in debates about human evolution. My recent visit to Lola illustrates what we have learned about this enthralling ape in the last couple of decades.

Walking around, Claudine and I converse in French about the challenge of maintaining a forested enclave close to the capital of the Democratic Republic of the Congo (DRC). It requires constant vigilance against people encroaching on the land, signs of which are easy to spot. Kinshasa, a megacity with poor infrastructure, has grown to at least 12 million inhabitants—no one knows the exact number. Despite the city's size, it is hard to hear its noise at Lola. The enclave truly is a paradise that contains around seventy-five bonobos rescued from the bushmeat trade.

In 1993 Claudine adopted a few starving bonobos from the city zoo and a medical lab. Before Lola was established, this beautiful area had a hotel with access to the nearby Lukaya River and waterfalls. Mobutu Seso Seko, the country's former president, would come here to relax on the weekends. His swimming pool now provides water storage for the bonobos.

Claudine talks passionately about the bonobos. She knows every single one by name and each one's heartbreaking backstory. Most had been clinging to their mother while she was shot out of the trees. Since the baby apes were too skinny to be sold as food and were thought to make cute pets, poachers brought them alive to the black market. Sometimes they were chained next to their mother, who at this point was a pile of meat. People would buy them, not realizing that they seldom survive without maternal love and the right nutrition. Since keeping a bonobo is against the law, orphans are regularly confiscated and brought to Lola. Some have been spotted among the baggage at the international airport, slated for illegal export. They arrive full of parasites, with bellies bloated from malnourishment, or covered with cigarette burns from years spent in a small cage at a bar. They often have raw red skin on their legs or neck from having been attached by a tight cord. At Lola, they are medically treated, dewormed, and bottle-fed at the nursery. Here they recover from their trauma and run happily around with their fellow orphans.

Upon its arrival, each baby ape is assigned a local woman who becomes their surrogate mother, or *maman* in French. This woman will spend all her time carrying, feeding, cleaning, and entertaining this one specific orphan. She will also scold them if they become too boisterous or annoy each other. Orphans stay in the nursery until they are about five, when they are transferred to one of the larger groups. It is easy to demonstrate how attached they become. Sitting next to a *maman* and her bonobo infant, Claudine invites the latter into her lap. The little ape does all the naughty things youngsters do, such as climbing all over Claudine, pulling her hair, trying to grab her glasses, and peeking under her clothes. At a hand signal by Claudine, the *maman* silently gets up and walks away. The bonobo immediately loses interest in her games, utters a sharp distress peep, and hurries after her *maman*. She has the pouting face of a frustrated bonobo, terrified of being left behind.

Claudine was only three years old when she arrived in the DRC, then called the Belgian Congo, where her father had been sent as a civil servant. She is charismatic, well connected, and highly respected in the country for her determination. She can get things done here that no one from the outside can accomplish. I am full of admiration for her project's scale, which is now under the direction of her daughter, Fanny Minesi, and her veterinary husband, Raphaël Belais. It requires professional fund-raising, administration, and a tight organization. Caring for all these apes, Lola employs a large Congolese staff to maintain the grounds, provide security, grow fruits and vegetables, and guide the public. Claudine knows not only all the bonobos but also all the people by name. Following Congolese custom, it's polite to address those over a certain age as *Papa* or *Mama* plus their first name, such as Papa Didier or Mama Yvonne. I am Professeur Frans. Everyone has a well-defined job and cares deeply about the sanctuary and its charges. Papa Stany, who has worked here since day one, has tears in his eyes when I inquire about Mimi, one of the first bonobos to be taken in.

This slender female with her long face and outsize ears arrived as an eighteen-year-old virgin. Coming from a pampered life in a home, she had watched television, flushed the toilet after use, played with children, leafed through magazines, eaten from the fridge, washed her hands, and slept in a real bed. No wonder Mimi tried to order people around once she had settled in at Lola. She didn't like to eat with those hairy savages (the other bonobos) and preferred to consume her fruits and vegetables on her own. She would order meals by clapping her hands. She had her little spot on a hill where she'd wait while all the other bonobos were being fed on the island. Papa Stany would throw her a huge bottle of milk, and no one would dare take it from her. As a bonobo with staff, she quickly became known as "Princess Mimi." But I've also heard comparisons with a queen or empress.

In the early years, Mimi had a habit of plotting escapes. One day she showed up at Claudine's door with a hammock and blanket under her arm, determined to return to the life of luxury that she had once known. After Claudine spent a few days indulging her, she was returned to the other bonobos. The next day, however, she fell dreadfully ill. She looked as if she were close to death, didn't react to kind words or caresses, and was barely able to lift her head. When the veterinarian arrived to treat her, he forgot to close the door behind himself (understandably, given the shape the patient was in), which created an opening for a renewed escape attempt. Mimi's health and energy returned instantly—and she was gone! Even though apes have a reputation as masters of deception, Mimi had still managed to trick everyone.[3]

I watched a film about her first meeting with members of her own species. The introduction went smoothly, as several females attempted to kiss her or presented their genitals to her. Mimi was not used to sexual advances, though, and had no clue how to react.[4] Bonobos engage in sex in every possible partner combination, and female-female sex carries special significance. It is the glue of their sisterhood. The most common pattern is *GG rubbing* (genito-genital rubbing), also called the *hoka-hoka*. One female wraps her arms and legs around the other and clings to her. While they face each other, the two press their vulvas and clitorises together, rubbing them sideways in a rapid rhythm. The bonobo clitoris is impressively long. Females carry big grins on their faces and squeal loudly during GG rubbing, leaving little doubt about whether apes know sexual pleasure.

Mimi, however, had no idea what the other females wanted. She also didn't understand what was up with the line of males trailing her wherever she went. Their erections were hard to miss because they eagerly showed them off—by sitting in front of her with their legs spread apart. Their long, thin pink penises, standing out against the dark hair on their bellies, made for an unmistakable signal. When males invited her this way, Mimi could only turn to her human caretakers with a questioning stare as if they could perhaps tell her what all the fuss was about.

With time, she learned to enjoy sex. As the first alpha female at Lola, she ruled the group with an iron fist surrounded by her female allies. Males who offended her by storming too closely past her or otherwise showing lack of respect could count on a severe beating. They'd be "corrected" by the central females, as Claudine likes to call it. It's the way bonobo females operate everywhere. Their dominance is not individual but collective.

Mimi's rule ended unexpectedly, however, when she gave birth to her first offspring. She died right afterward. Papa Stany, who had been Mimi's primary caretaker, was devastated. Her passing came as a shock to everyone. This sad event occurred ten years before I visited Lola, but the immense love for Princess Mimi was still palpable.

Mimi's initial aversion to sex had a male parallel. A full-grown bonobo named Max came to Lola after spending many years among gorillas at a sanctuary in Brazzaville. He became known as "the gorilla" because of the guttural grunts he produced during meals. Gorillas utter a constant stream of deep food-grunts while they chew their vegetables, also known as "singing" or "humming."[5] In contrast, bonobos utter high peeps. Being used to gorillas, Max sang like them. He also failed to develop an appetite for ballooning genitals, which aren't part of gorilla anatomy. Despite his popularity with the bonobo females, Max ignored their courtship. Semendwa, who took over as alpha after Mimi's death, didn't give up, though. She'd stare into Max's face, then at his flaccid penis, back and forth, trying to figure out what could be the matter. She'd tickle his testicles with her fingers to see if that might do the job, but it didn't.

It took Max a long time to turn into a real bonobo.

Princess mimi reminded me of Prince Chim, another legendary ape. Prince Chim was thought to be a chimpanzee, but the American ape expert Robert Yerkes felt that he differed from every other ape he knew. Chim had an admirable personality and showed special concern for his terminally ill female companion. In 1925 Yerkes wrote, "I have never met an animal the equal of Prince Chim in approach to physical perfection, alertness, adaptability, and agreeableness of disposition."[6] A postmortem inspection of this ape concluded that he was a bonobo.

Bonobos were recognized relatively late as a species. Only in 1929 were they distinguished from chimpanzees based on their anatomy. The original name was *pygmy chimpanzee,* but this name exaggerated the size difference. Chimpanzees look as if they work out in the gym every day. They have large heads, thick necks, and broad, muscular shoulders. In comparison, bonobos have an intellectual look, as if they spend their time in the library. They have slim upper bodies, narrow shoulders, thin necks, and elegant piano-player hands. A lot of their weight is in their legs, which are long and thin. When a chimpanzee knuckle-walks on all fours, his back slopes down from his powerful shoulders. A bonobo, in contrast, has a perfectly horizontal back because of his elevated hips. When standing on two legs, bonobos straighten their back and hips better than any other ape so that they look eerily human-like. They walk upright with remarkable ease while carrying food or looking out over tall grass. Of all the great apes, the bonobo's anatomy is closest to that of Lucy, our *Australopithecus* ancestor named after a 3.5-foot, 4-million-year-old juvenile female fossil.[7]

Apes are sometimes called quadrupedal (four-footed), but bonobos are quadrumanual (four-handed). Their hands and feet are fully interchangeable. They may use a foot to pick up something, hold an object or an infant, kick each other, masturbate, or reach out for contact.[8] The universal hominid begging gesture, with outstretched open palm, is often assigned to a foot if a bonobo has her hands full. Bonobos jump, brachiate, and flip around in the trees with unbelievable agility. High above the ground, they cross lianas on two legs like fearless tightrope walkers. These apes have never been forced out of the forest and hence never needed to compromise their tree-dwelling ways.

That bonobos are more arboreal than chimps is evident when they encounter strange people in the woods. Suehisa Kuroda, a Japanese primatologist, studied bonobos, which typically escape through the canopy and descend to the forest floor only once they're far away. Then he went to see wild chimpanzees and had to get used to them dropping out of the trees and fleeing over the ground, so different from the bonobos he was used to. Kuroda was shocked by chimps scattering in all directions. Even mothers and their young might take different routes. Bonobos would never do such a thing. They stick together.

Bonobos still live in the original swampy rain forest where apes likely evolved. For this reason, they may most resemble the original apes from which all African hominids descend, including humans. This ancestor may also have shown the arrested development that marks both bonobos and us. Our species is *neotenous,* which means that we carry fetal or juvenile traits forward into adulthood. Examples of neoteny are our naked skin, bulging cranium, flat face, and frontally oriented vulva. We also retain the playfulness and curiosity of juveniles. We play, dance, and sing until we die, and we keep exploring new knowledge by reading nonfiction or taking senior classes. Neoteny has been called the hallmark of our species.[9]

Bonobos have sipped from the same youth potion. They, too, stay forever young.[10] All their life, they retain the cute white tail-tufts that chimpanzees lose upon weaning. Adult bonobos possess the small rounded skulls of infant apes and remain remarkably playful. In most primates, adult males are more playful than adult females, but not in bonobos. It is not uncommon to see bonobo females frolicking around, tickling and chasing each other with hoarse laughing sounds. The species has other neotenous traits, such as their more open faces that lack the other apes' prominent eyebrow ridges. They also share with us a frontal vulva with a prominent clitoris, which make face-to-face copulation and GG rubbing favorite positions.[11]

The most juvenile trait of all, however, is their high-pitched voice. The easiest way to tell chimps and bonobos apart is by ear. The drawn-out *huu-huu* hooting of the chimpanzee is absent in the bonobo. Adult bonobos of both sexes have such shrill voices that at first you think you are hearing a monkey or young ape. Since bonobos are only slightly smaller than chimps, their high timbre is not due to their body size. Rather, they have a modified larynx. Perhaps they sound like juveniles because in their society they have less need for intimidation.[12]

In the 1930s, the Hellabrunn Zoo in Munich received a shipment of bonobos from Africa. The director, who had not yet looked under the cloth covering the crates, almost sent them back. He refused to believe that the sounds he was hearing were coming from the apes he had ordered. The Hellabrunn bonobos appeared in the very first behavioral study of the species. Eduard Tratz and Heinz Heck published their findings after the war, in 1954. They drew up a list of differences between bonobos and chimpanzees, including the bonobo's sexual behavior and gentle nature. To describe their sexual habits, they resorted to Latin, saying that chimpanzees *copula more canum* while bonobos *copula more hominum* (chimps mate like dogs and bonobos like humans). Echoing Yerkes's opinion, they concluded, "The bonobo is an extraordinarily sensitive, gentle creature, far removed from the demoniacal *Urkraft* [primitive force] of the adult chimpanzee."[13]

Sadly, the Hellabrunn bonobos died on the night in 1944 when the World War II allies bombed Munich. Terrified by the noise, they all succumbed to heart failure. That none of the zoo's other apes suffered the same fate attests to the bonobo's exceptional sensitivity.

I saw bonobos for the first time at a now-defunct Dutch zoo that kept a couple of what it labeled "pygmy chimpanzees." However, they seemed too different in physique, demeanor, and behavior to qualify as such. They were also not that small: bonobos are the same size as the smallest subspecies of chimpanzee. Because we knew virtually nothing about them at the time, I decided that this had to change. A good starting point, I thought, would be to get rid of the "pygmy" label. It was misleading and demeaning, as if they were the poor man's miniature chimp. People would ask, "Why study those little chimps if you can study the real big ones?" I agreed with Tratz and Heck that bonobos deserved their own name. We don't know the origin of their name, but one speculation is that it derives from a misspelling on a shipping crate from Bolobo, a town in the DRC. Regardless, I made a point of always calling these apes "bonobos" despite the resistance of journal editors and blank stares from the general public. The new name took hold owing to its happy ring, which befits the species' nature.

At the same Dutch zoo visit, I watched a minor squabble over a cardboard box. A male and female ran around and punched each other, but all of a sudden, their fight was over. They were making love! This was an odd turn of events: chimpanzees don't switch so quickly from anger to sex. I thought the change of heart was a coincidence or that I had missed something that could explain it. In retrospect, however, there was nothing unusual about what I had witnessed.

Today, we know more about the genetic background of our two closest relatives. According to DNA analyses, we have no reason to favor one over the other for comparisons with us. We share some genes with bonobos that we don't share with chimps, but we also share some genes with chimps that we don't share with bonobos. Genetically, both apes are exactly equally close to us.[14] The split between them and us occurred between 6 and 8 million years ago, but indications are that it was a long and messy divorce. While our ancestors charted their own path, they kept coming back for trysts with the apes. Human and ape DNA show signs of a million-year-long hybridization phase, not unlike the continued interbreeding today between grizzlies and polar bears or wolves and coyotes.[15]

What happened 6 million years ago matters for the story of human evolution. Traditionally, we assume that our ape forebears looked and acted like today's chimpanzees. This is pure speculation, though. Fossilization in the forest is so poor that our ancestral hominid remains a mystery.

All three survivors—bonobos, chimpanzees, and humans—have evolved since then. No species stands still in time. It's a mere historical accident that early explorers ran into chimpanzees first, which is why science still turns to them when discussing our pedigree. If the explorers had run into bonobos first, these apes would now be our primary model. Think of the fascinating implications that might have had for our ideas about gender!

Given how much we share with bonobos, including our celebrated neoteny, the thought that we descend from a bonobo-like ape is not far-fetched. After all, Harold Coolidge, the American anatomist who gave the bonobo its species status, concluded from his dissection of Prince Chim's corpse that this ape "may approach more closely to the common ancestor of chimpanzees and man than does any living chimpanzee." A recent anatomical comparison arrived at the same conclusion.[16]

Two male bonobos rub their rumps together. These contacts are less common and less intense than the GG rubbing among females.

Mʏ 2019 sᴛᴀʏ at Lola gave me a refresher course in bonobos. I hadn't worked directly with them since the 1980s. At the time, we had the admirable field studies by the Japanese primatologist Takayoshi Kano, who would go on to offer the first outline of this ape's society in his 1992 book *The Last Ape*. We had language studies with Kanzi, an ape genius who had learned the meaning of a large number of lexigrams. And we had my own work on communication and sexual behavior among bonobos at the San Diego Zoo. But this was about all the bonobo research that was going on in those early days.[17]

Since then, much has happened. For a decade, political turmoil and a horrible war in the DRC interrupted fieldwork there, but it is now back in full force. Studies of bonobos in captivity, including experiments on intelligence, have also taken off. And my own team explores bonobo empathy by documenting their soothing reactions to the distress of others. This work at Lola is being led by my longtime collaborator Zanna Clay, a professor at the University of Durham in the UK. I am here to see Zanna and discuss our project as well as renew my acquaintance with bonobos.[18]

While I have always loved these fascinating apes, and can't get enough of contrasting them with their more robust sister species, the early days of discovery were far from easy. The world of science was ill at ease with bonobos and their behavior. Accepting them as next of kin undermined the way we looked at ourselves. Only a handful of scientists knew firsthand how unique bonobos were, but we had a hard time getting our message across. Bonobos were too sexy, too peaceful, and too female-dominant to please everyone. They made some visibly upset, as once when I lectured to a German audience about the power of bonobo alpha females. Afterward an older male professor stood up and barked in an almost accusatory tone, "What's wrong with those males?!"

Since apes hold up a mirror to ourselves, we care how they make us look. Perhaps the biggest problem with bonobos was their nonviolence. We have no confirmed reports of one bonobo killing another, whereas we have an abundance of such cases for chimpanzees. You'd think everyone would be pleased to get a break from chimpanzee brutality and finally meet a close relative leaning toward love rather than hate. But then you wouldn't have reckoned with the prevailing narrative in anthropology, according to which we are born warriors who conquered the earth by eliminating every ancestral type that stood in our way. We are children of Cain, not of Abel.[19]

These views date back to a 1924 fossil find in South Africa. Dubbed *Australopithecus africanus*, this forebear was portrayed as a carnivore who swallowed his prey alive, dismembered them limb from limb, and slaked his thirst with their warm blood. The paleoanthropologist Raymond Dart came up with this vivid description despite having only a single juvenile skull to work with. The scant evidence couldn't keep his imagination from running wild. We now realize that *Australopithecus,* who looked a lot like an upright bonobo, was nowhere near the top of the food chain. Still, Dart's gruesome characterization has stayed with us. It inspired the "killer ape" myth, according to which we descend from merciless murderers and rapists who waged war almost for fun.[20] When chimpanzees' violent nature became more widely known, this clinched the theory. With similar tendencies attributed to both our ancestors and our ape relatives, who could doubt that we have a taste for blood in our heritage?

These ideas worked out to everyone's satisfaction—until the pacific bonobo burst onto the scene. According to Takayoshi Kano, groups of them meet in the forest without any fighting. His students even spoke of "mingling" and "fusions."[21] Today we know that bonobos share food between communities and occasionally adopt orphaned youngsters from their neighbors. These field reports have thrown a huge wrench into the accepted human origin myth. My own studies, elaborating the erotic and hedonic side of the species, made matters even worse. Bonobos became polyamorous flower children. Having such a sweet, sensual character in the family didn't jibe with the assumption of unbridled violence throughout human prehistory.

The dominant hypothesis remains that we carry the mark of Cain. For example, in his 2011 book *The Better Angels of Our Nature*, the Canadian-American psycholinguist Steven Pinker proposed that humanity needs civilization to keep its destructive instincts under control. Since his theory works only if our forebears were hyperaggressive characters, Pinker went for the chimpanzee as ancestral model and cheerfully swept bonobos under the rug, calling them "very strange primates." In the same vein, the British-American anthropologist Richard Wrangham in his 2019 book *The Goodness Paradox* concluded that humans are better at living together than you'd expect, so we must have domesticated ourselves. He too uses an aggressive chimpanzee-like ancestor as his starting point, whereas bonobos are an evolutionary offshoot who "have gone their separate way."[22]

The inconvenience of bonobos in our family tree is on full display in these books. Never mind that both Pinker's and Wrangham's evolutionary scenarios would be unnecessary if our species hailed from less belligerent stock. Had we descended from a bonobo-like ancestor, things would be much simpler. No special explanation would be required for our species's moderate levels of violence. Instead of posing a problem, bonobos might be the solution.

A second sensitive issue about bonobos was their sex life. It posed a problem due to the hang-ups of some human cultures. Nature documentaries by prominent international outlets, such as the BBC and Japan's NHK, didn't want to touch the issue of sex with a ten-foot pole. They'd show footage of grooming and frolicking bonobos, but they'd freeze the image as soon as the apes adopted positions in which something sexual was imminent. The narrator would lead viewers astray with some vague remark, such as that bonobos enjoy their time together. I dubbed it the *coitus interruptus* treatment.

Scientists were troubled as well. One wrote that we're better off ignoring these "weird" apes whose X-rated sex life "sounds exhausting." Another tried to question the high rate of bonobo sex. His calculations, however, were limited to adult heterosexual encounters, thus leaving out a huge chunk of bonobo erotic activity. Some colleagues even refused to recognize the sexual nature of genital stroking and rubbing. "Is it really sex?" they'd ask. They preferred to label it extreme affection. This was almost funny! I couldn't help but point out that if I were to show this kind of "affection" on a busy street, I'd be in handcuffs within minutes.[23]

Frans Lanting, a famous wildlife photographer, approached me with thousands of pictures of bonobos taken during a *National Geographic* expedition to the DRC. Most of the images had never seen the light of day as the magazine had deemed them too graphic. When I saw his treasure trove of fantastic shots, taken under the most trying circumstances (for a photographer, there is nothing worse than black subjects in a dark forest), I realized that they presented a momentous opportunity. As Dutchmen of the same age living in America, Frans and I had an easy rapport and decided to work together to raise public awareness. The explicit pictures in our 1997 book *The Forgotten Ape* have, to my knowledge, never bothered anyone.[24]

The third and final contentious issue about bonobos concerned the relation between the sexes. All scenarios of human evolution did and still do assume male superiority. Female rule in a close relative undermines this narrative. I received the first hint of bonobos' unconventional social order when I studied them at the San Diego Zoo. Originally Vernon, an adult male, was housed with Loretta, an adult female whom he clearly dominated. But when Louise, an older female, was added to their group, the two females began to boss over Vernon. Vernon had to beg them for a share of food. I found this odd since he was a muscular male who was larger than the females and possessed the sharp canine teeth of his sex. However, as I came to know more zoo bonobo groups, I found female dominance to be the rule. In fact, I don't know of any bonobo colony that is led by a male.

Fieldworkers suspected the same but were reluctant to make such a bold claim. Then in 1992, at a congress of the International Primatological Society, investigators of both captive and wild bonobos presented data that left little doubt. The American anthropologist Amy

Parish reported on food competition in small zoo groups of chimpanzees and bonobos. A dominant male chimpanzee will claim the available food right away and consume it at his leisure while the females wait. In bonobos, in contrast, the females will be the first to approach the food. After some GG rubbing, they'll feed together, taking turns. Males may make as many charging displays as they like, but the females will ignore the fuss.[25]

At the same conference, fieldworkers confirmed female dominance. For example, when sugarcane was laid out in Wamba forest, in the DRC, the male bonobos arrived first and ate in a hurry, because once the females arrived, they would take over. All the males could do at this point was cram their hands and feet full of stalks and take off. Some scientists have questioned if this counts as dominance, suggesting that male bonobos are perhaps "chivalrous" in relation to food. This interpretation might be believable if the males just gave way, but this is not how things go. Females actively chase them off and sometimes attack them. The standard criterion, applied to every animal on the planet, is that if individual A can chase B away from its food, A must be dominant.

Kano answered these skeptics as follows: "Priority of access to food is an important function of dominance. Since most dominance interactions and virtually all agonistic episodes [conflicts] between adult females and males occur in feeding contexts, I find much less meaning in dominance occurring in the non-feeding context. Moreover, there is no difference."[26]

Takeshi Furuichi, one of Kano's students, reports that lone females at Wamba sometimes avoid a male who performs a branch-dragging display. Under these circumstances, the worked-up male is temporarily dominant. This doesn't mean, however, that he can attack the female or claim her food. When females are together, which they almost always are, they are confidently in charge.[27]

Could the female dominance at Wamba have been a product of the extra food that the investigators handed out? After all, this artificial situation incites competition. The problem with this explanation is that competition rarely alters the hierarchy—it only makes it more visible. We can see this in wild chimpanzees, in which females never dominate the feeding sites set up by fieldworkers. So the fact that bonobo females do so tells us something about their society.

At another field site, the LuiKotale forest in Salonga National Park, scientists have followed wild bonobos for twenty years without any food provisioning. Recently, they drew up a hierarchy among these apes based on confrontations and acts of submission recorded in the forest. The six top ranks in this hierarchy were firmly in female hands.[28]

At LOLA, the bonobos are fed from a little boat steered by Papa Stany, also known as *le Capitaine*. While I sit behind him to photograph the scene, the apes wade waist-deep into the water to pick up the papayas, oranges, and sweet potatoes that failed to hit land. Since bonobos can't swim, this is tricky. Several individuals, before they walk on two legs out into the lake, pick up long branches to assess its depth. They probe as they go. Both sexes do so, but I wonder if the same rule applies as in chimpanzees, in which females are generally the better tool-users.

It has been a puzzle why bonobos don't use tools in the wild, whereas chimps do so all the time. Could it be a difference in mental capacity, as some have suggested? Since the Lola bonobos are skilled tool users, it is more likely that their wild counterparts just don't need tools to find food.[29]

An excellent illustration is an incident Zanna Clay filmed while she was following Lisala. Lisala picked up an enormous fifteen-pound rock and heaved it onto her back. It was a surprising thing to do, but Zanna knew she was going to use that rock. It was a bit like when you see a man walking in the street with a ladder: he wouldn't be carrying such a load for no reason. Lisala went on a fifteen-minute stroll with the rock on her shoulders while her baby clung to her lower back. Along the way, she picked up a handful of palm nuts. Upon reaching a sizable rocky surface (the only one in the enclosure), she put down the rock, her infant, and the nuts. Then she began to crack the extremely tough nuts, placing them one by one on the anvil to pound them with her big rock. It's hard to imagine that Lisala had gone to all this trouble without a plan. By picking up her tool long before she could put it to use, and before she even had any nuts in hand, she showed the sort of forward thinking that has now been confirmed by experiments on apes.[30]

While we are feeding the apes, the tightness of the female community is evident. Females groom each other and engage in sex, and after we have shown them that our buckets are empty, they take off together into the forest. I call it a *secondary sisterhood* because their solidarity doesn't rest on kinship. In the wild, males stay in their native community all their lives, whereas females emigrate when they reach puberty. This means that females join neighboring communities in which they have no or few relatives. They form bonds with senior resident females whom they didn't know before. The same occurs at Lola, where females that arrived

Lisala, a bonobo, carries a heavy rock (and her baby) on the way to a place where she expects to find nuts. Once she has collected the nuts, she will use the rock as a hammer to crack them. Picking up a tool so long in advance suggests planning, a well-established capacity in the apes.

as orphans from various parts of the country band together despite a lack of family ties.

Bonobo female alliances are so strong that even human males notice them. I know several men scientists who tried to work with bonobos in captivity and ran into trouble because of uncooperative females. Bonobo females work better with female experimenters or observers. When Amy Parish studied the San Diego Zoo bonobos, the females embraced her as one of their own, something they'd never done with me. True, Loretta often solicited me from across the moat (turning her genitals to me while peeking between her legs and waving a hand at me), but this was purely sexual. She has always flirted with me and still does so whenever she sees me on a visit. But being a male, I was never part of the gynarchy that is a bonobo society. Amy, in contrast, once got food tossed at her from across the moat. The bonobos must have thought she was hungry.

In all primates, females bond over offspring. They do so partly for practical reasons because the young ones need playmates. It is common for mothers to seek out other mothers with similar-aged offspring. While they groom each other, the young ones wrestle and race around under their eyes. When Amy visited her old bonobo friends, who had been moved to another zoo, she wanted them to meet her newborn son. The bonobos recognized her right away. The oldest female briefly glanced at Amy's baby from across the moat but then ran indoors. She quickly returned with her own baby, whom she held so that the two infants could look into each other's eyes.

The strong alliance among central females in a bonobo group, such as the one that surrounded Princess Mimi, is not necessarily nice. There is a widespread assumption that female dominance must be less harsh than that of males. When the journalist Natalie Angier summarized bonobo society in *The New York Times,* she softened the status of females: "the dominance is so mild and unobnoxious that some researchers view bonobo society as a matter of 'co-dominance,' or equality between the sexes."[31] Perhaps this is what we believed in 1997, but hierarchies always entail coercion. This is as true for females as it is for males.

Alpha females typically reach their position based on age and personality. Since these traits are unalterable, challenges are rare. This is why female hierarchies are usually more stable than those of the males. But females, too, occasionally need to remind others who's in charge. At Lola, I saw Semendwa grab a low-ranking female foot and deliver a firm bite that drew blood. This female had committed the faux pas of approaching a papaya on which Semendwa had laid her eyes. The screaming victim was lucky that bonobos rarely inflict worrisome injuries. It was just a little cut. Still, she got a painful reminder not to cross the alpha female.

Dominant females operate more harshly against males who don't respect their priority around food or provoke them by displaying too closely. Given their speed and agility, males often get away, but things may turn ugly if they do get caught. At Lola, this sometimes happens in the night quarters. In the evening, the whole group enters a building for a night's rest. If a male is cornered there, the females may take off a finger or even go for his testicles. Males learn to be cautious. Most of them enter the building last and are the first to be released next morning. The exceptions are males with strong ties to the females.

At zoos, I invariably hear about management problems with bonobo males. They are hard to integrate due to female aggression. As a result, zoos keep them separate from the females most of the time. The good news is that, with better information on the species's natural behavior, we now know how to avoid these problems. Bonobo males are mama's boys in that they rely on maternal protection. In the wild, a son will continually keep his mother in sight.[32] Her presence deters other females from turning against him. Mother-son combinations sometimes act as power couples with mutual benefits, especially if the son is attractive to other females. For zoos, this means that sons should always be kept with their mothers and not be moved around independently. Now that they keep this rule in mind, things have improved enormously.

In the natural habitat, social tensions may be uncommon, but they aren't absent. For example, in Lomako Forest in the DRC, an adult male bonobo made a threatening move on a low-ranking female with a newborn. She almost lost her balance in the tree but then pushed him off her branch and pursued him with shrill screams. Fifteen or more apes joined in a fierce attack against this male. The outbreak of violence suggests that bonobo society has a deeper layer, one generally hidden by its Woodstock facade. Other field studies confirm coordinated female protests against male harassment. Thanks to their camaraderie, females keep violent males in check. Their solidarity even crosses group boundaries. When groups mingle in the forest, females from different groups may band together against aggressive males.[33]

I sometimes ask my scientist colleagues if they feel that bonobo males have good lives. They look bewildered, since this is not your typical science question. We don't have theories about which organisms have good or bad lives. I am used to talking with zoo curators, though, who worry about their male bonobos. And for some men, such as the German professor I met, being dominated by women is about the worst thing they can imagine. This is why I want to hear my colleagues' evaluation of bonobo males' quality of life.

For captive bonobos, the scientists tell me, males' quality of life varies with the colony's size and the amount of available space. In colonies with little space, serious friction arises, of which the unlucky males bear the brunt. Bonobos in zoos with large forested outdoor enclosures fare much better, such as in Apenheul in the Netherlands or La vallée des singes in France. In those colonies, the males are doing very well.

But what about bonobos in their natural habitat? After all, this is where bonobo society evolved. Here, my colleagues explain, males have few worries. They stay out of trouble by regulating their distance to the core of the group. They hang out with the females if all goes well, but if things get edgy, it's easy for them to get away. They just disappear for a while. Most of them are well liked and enjoy plenty of sex and grooming with the females. They are an integrated part of the community.

Male bonobos generally lead long lives. Their risk of injury and death is lower than for male chimpanzees. Chimps kill between groups, and sometimes even within their group. Their status struggles can get incredibly tense. When they fight, the damage is far worse than in bonobos. Male chimps' attacks on females are rarely life-threatening, but they are still rough and abusive. So both sexes have a great deal of stress to cope with. After a lifetime of fieldwork on both ape species, Furuichi and his wife wondered how it would feel to be like them: "This is why I say, 'I do not want to be a chimpanzee male,' and, in response, Chie Hashimoto, my wife, tells me, 'I do not want to be a female chimpanzee.' "[34]

Lola is more than a sanctuary for rescued apes. It welcomes many visitors and school classes from the city to teach students about bonobos and the need to protect them. Getting the conservation message out is crucial in a country so rich in flora and fauna. Being four times the size of France, the DRC has vast stretches of rainforest to preserve. Claudine has spoken to thousands of people and regularly appears on national television. If the bonobo is well known to the Congolese people, it's thanks to her.

Lola is actively engaged in conservation. It's one of the few sanctuaries that has successfully reintroduced primates to the wild. This is no easy task, as failures can occur for many reasons. Animals released from sanctuaries have less disease resistance. They can't compete with wild residents of their species. They lack knowledge about natural foods and dangers. And they don't know how to fend for themselves.[35]

At Lola, however, bonobos have a natural tropical forest as their training ground. They learn about possible risks, such as venomous snakes. They learn which plants and fruits are good to eat, and which ones make them sick. Moreover, once they are released in a wild forest, bonobos run fewer risks due to hostile residents of their species, because these apes are far less xenophobic than most primates.

Twice already, Lola has returned a group of bonobos to a natural habitat. Shipped and flown one thousand miles to the north, they were released in a protected area that now covers 120,000 acres of primary forest called Ekolo ya Bonobo (Land of the Bonobos, in Lingala). These lucky bonobos moved from the Lola nursery to survival in the wild! Closely monitored by observers, the released apes are making it on their own. They feed themselves without human help and have produced five babies since their release. The reintroductions have been a great success.

It's a genuine accomplishment for Claudine and her daughter. Claudine, who is close to retirement, described to me her vision for Lola and its release program. She stressed the role of the local human population. Conservation is not just about animals, she said; it's even more about people. When the people are on your side, everything is possible, so community projects have been set up around Ekolo. Now every time Claudine arrives by boat (rivers are the roads of the DRC), the villagers come out in their best clothes to dance and sing on the shore.

We also discuss the conspicuous role of women in the sanctuary movement.[36] Lola is the world's only one for bonobos, but there are many sanctuaries and rehabilitation centers in Africa for chimpanzees, gorillas, elephants, rhinos, and other wildlife. Virtually all of them were founded and are run by women. This is also true, by the way, of sanctuaries for ex-laboratory or ex-pet primates in the West. Even the well-known David Sheldrick Wildlife Trust, despite its name, was founded by a woman. Daphne Sheldrick named the sanctuary after her late husband. While he was busy establishing a large national park in Kenya and battling ivory poachers, she adopted and bottle-reared hundreds of orphaned elephant calves. The overwhelming female involvement in sanctuaries reflects a protective role that we also recognize in the American pioneer of ecological engagement, Rachel Carson, and in today's environmental crusaders, from Jane Goodall to Greta Thunberg.

Some conservationists look down on the sanctuary movement. They prefer to tackle bigger issues, such as battling the logging companies and preserving entire ecosystems. This is crucial, but we can't just turn our backs on young bonobos who were ripped from their mothers' arms and are crying out in distress. I am immensely grateful for people in the world, such as Claudine, with the heart to care. We need to protect vulnerable individuals as well as the health of the planet.

There's no reason why we can't do both.

6

SEXUAL SIGNALS

From Genitals to Faces to Beauty

The flamboyantly colored face of the male mandrill echoes his derrière. A red line along the middle of his face flanked by blue paranasal ridges replicates his bright red penis against blue buttocks. Even his orange goatee copies the orange tufts of hair below his scrotum.

Similarly, the female gelada baboon repeats on her chest the patterning of her rump: her two bright red nipples are placed so close together that they look like labia. The naked skin around them resembles that on her behind. We wonder about the function of these showy signals of monkeys and smile at their weird body self-mimicry.

But might the same apply to us? In 1967 Desmond Morris speculated in *The Naked Ape* that a similar back-to-front signal migration had befallen our lineage. Our red lips mimic a vulva. Women's breasts have the rounded shape of buttocks. A man's bulbous nose recalls a flaccid penis. Not everyone was amused: commentators slammed his book as "salacious guesswork." I have nothing against calling Morris's theories wild or unsupported, but do we really still need to throw a Victorian hissy fit when it comes to genitalia? It's not as if these body parts leave us cold. We find them irresistible! Look at bronze statues, such as the anatomically correct Charging Bull in New York's financial district. Or the statue of Victor Noir in Paris, famous for the prominent bump in his trousers. The polished parts of these statues betray the rubbing of genital areas by thousands of eager human hands. Michelangelo's *David* is lucky to stand so high above the crowd.

Given the trouble we have agreeing on behavior, let alone explaining it, anatomy offers a perfect launching pad for a discussion of human biology. Morris's speculations may seem outrageous, but the questions won't go away. Why are we the only primate with everted (turned inside out) lips, which makes them contrast with the skin around them? Lips don't serve as sexual signals in other primates, so why do females of our species so often enhance them with lipstick and slightly part and lick them in suggestive ways? Why are we the only primate with permanently protruding breasts, often pushed up with the help of bras or injected with silicone? Breasts don't need this shape for effective nursing. Why do we have a pointy nose that sticks out, whereas other primates have no trouble smelling without such an odd contraption on their face? For the evolutionary biologist, these are valid questions.

Morris's tongue-in-cheek style took the sting out of a hugely sensitive topic at a time when even the word *naked* was deemed naughty. His book had serious undertones, however, such as being the first to explicitly attack the *tabula rasa* view, according to which we arrive on earth like an empty sheet on which the environment writes anything it likes. Morris forcefully rejected this pre-Darwinian notion and in so doing paved the way for popular writers about evolution such as E. O. Wilson, Stephen Jay Gould, Richard Dawkins, and others. But the main reason for his book's huge success—it is still the only biology book that ranks among the one hundred most-read books in the world—was that it poked fun at our species while wobbling its pedestal. It gave readers a combination of surprising observations and good laughs.

Referring to *Homo sapiens*, Morris offered this gem: "This unusual and highly successful species spends a great deal of time examining his

higher motives and an equal amount of time ignoring his fundamental
ones. He is proud that he has the biggest brain of all the primates, but
tries to conceal the fact that he also has the biggest penis."[1]

Morris wrote these words well before we knew much about
bonobos. This ape's long penis makes the majority of men look under-
size, all the more so after correction for the bonobo's smaller body. Their
pink penis is thinner than that of humans, though, and fully retractable.
Its color makes an erection rather eye-catching, especially if the male
flicks it up and down. Even more remarkable than the ability to "wave"
his penis is the fact that the bonobo's testicles are many times human
size. This relates to the amount of sperm required when females mate
with multiple partners. To have any chance at achieving fertilization
amid the loads from other males, a male needs to send a massive stream
of one-celled swimmers toward the egg.

Every time I hear about *manspreading*—a term that entered the
Oxford English Dictionary only in 2015—I can't help but think of primate males showing off their genitals. Women complain about the space
men take up by sitting with their legs apart on public transportation.
This unconscious male posture is often attributed to socialization and
male entitlement, but it's universal among primates. For example, if you
walk behind a male vervet monkey, you won't be able to miss his bright
blue testicles, but they also stand out frontally when he sits down with
his legs apart. Male primates often sit like this, as if everyone needs to
know what sex they are. They also adopt this posture while soliciting a
female. By showing off a stiff penis, they signal both eagerness and ability to perform.

Wide-legged postures convey dominance and also serve as a threat.
In squirrel monkeys, a male may thrust his erect penis into the face of
a cowering subordinate, who ducks out of the way. Only high-status
males dare display their genitals. If you encounter a monkey troop with
a male sitting with opened legs in full view of everyone, you can be sure

The ancient Egyptians worshipped baboons. Known as aggressive and virile, their statues of these monkeys empha-sized male genitals.

that he is at the top of the social ladder. Given the vulnerability of these body parts, it takes self-confidence to flaunt them. Subordinate males watch their backs as well as their bottoms. They try not to draw atten-tion, and they shroud their sexual interests in secrecy.[2]

The connection between dominance and penile displays was well known to the ancient Egyptians, who depicted their sacred baboons as males sitting with legs apart, hands on their knees, and penis visible. The same connection exists in the gigantic phallic symbols of power and victory in our societies, from the Washington Monument to the Eiffel Tower. Even our insults resemble penile signals, such as when we raise a middle finger or jerk a forearm upward while holding the upper arm still with the other hand. The finger gesture was already known to the ancient Greeks and Romans, who called it the *digitus impudicus* (indecent finger) in Latin.[3]

Obviously, none of the above is an excuse for men to take up more space than they need on the subway. While a woman looking for a seat may consider manspreading appalling, researchers have conducted an actual study to see if women might find this posture appealing under other circumstances. Using a speed dating application, the American psychologist Tanya Vacharkulksemsuk found that "postural expansion" works for men. Pictures of men holding a power pose with spread limbs and a stretched torso were contrasted with men adopting folded postures. Poses that take up space convey openness and dominance, which helped men achieve a romantic connection. In the study, few of the men passed muster with the women, but those who did almost always held an expansive pose.[4]

Attention to male genitals reflects a general focus on male sexuality at the expense of that of females, who are assigned a passive role. Females are often considered the recipients of sex, not the seekers. Attitudes are changing, however, even in biology. I could offer a variety of animal examples, but let's stay with our nearest kin. Female apes are active players who often try to have sex with a great variety of males. Why do they do so if one male would suffice to achieve pregnancy? Why not pick the best available male and leave it at that? And why do so many women do the same? Morris didn't answer this question when he depicted human evolution as revolving around male hunters while delegating females to a childbearing role.[5]

The myth persists that evolution takes place mostly through the male line. Open any book on human prehistory, and you'll see images of men waging war, making fire, hunting big game, building huts, and defending women and children, who are fearfully huddling together against outside threats. These scenes may well have happened, but why are men always the heroes of the story? Didn't women contribute to the success of our species? The most outrageous statement in this regard (and there are many to choose from) came from the American surgeon Edgar Berman, who in *The Compleat Chauvinist* boasted, "We males have been born the fittest for three billion years."[6]

I'm afraid that this remark made Berman sound like a "compleat" idiot. The concept of evolutionary fitness is not to be confused with common usage of the term to mean individual physical fitness, as in "fitness exercise." It's not about who can jump the highest or run the fastest. Fitness is defined in biology as success in survival and breeding. It may come about through a superior immune system, keener eyesight, better camouflage, larger lungs, or any other beneficial trait. Since fitness is measured by one's genetic contribution to the next generation, it is logically impossible for the members of one sex as a whole to be fitter than those of the other sex as a whole. Fitness is indivisible. In sexually reproducing organisms, mothers and fathers make equal distributions to the genome. If the males of a species fare poorly, the females will go under with them. Conversely, if the females fare poorly, the males can forget their genetic legacy. One sex being fitter than the other would be like a galley that puts all its strong rowers on one side of the ship, and all its weak ones on the other side. It would be going around in circles.

FEMALE FITNESS HAS distinct requirements. True, both sexes need to eat and stay out of the claws of predators, but they contribute differently to the next generation. You'd expect females to pursue their own agenda rather than being resigned to their fate. The enterprising sexuality of females, known as female choice, has become one of the hottest topics in biology. It's also called female promiscuity, but this word has too much of a moral ring to it, and a negative one at that. I prefer to call it female sexual adventurism or proactivity. This phenomenon used to be an enormous taboo, as if females can be only faithful, coy, and choosy.

The growing evidence for female sexual adventurism has shifted the focus from the penis to the clitoris and from the male sex drive to the female orgasm. Female empowerment has reached evolutionary biology.

At one time even the presence of a clitoris in our fellow primates was in doubt. If one was found, it was confused with a penis. One nineteenth-century report spoke of a "hermaphrodite orang outang" but featured a gravure of a gibbon known for its penis-like clitoris. A famous eighteenth-century monkey in the Royal Museum of Physics and Natural History in Florence, Italy, was considered to be a hermaphrodite. Experts fought over the status of this "monstrosity," which was said to make museum visitors blush. All this because some primates sport a clitoris so large that a female may be mistaken for a male.[7] This is especially true for neotropical primates. In our capuchin monkey colony, for example, we once welcomed the birth of a young male with prominent genitals whom we named Lance. Years later Lance's behavior became increasingly odd. Our suspicions were confirmed when a chromosomal test showed "him" to be a her.

Another neotropical primate well known for its elongated clitoris is the spider monkey. My longtime Italian collaborator Filippo Aureli and I, standing in a Yucatán forest in Mexico, peered up into the trees with binoculars to spot these monkeys. Filippo's study subjects climbed around high above us, which made it hard to make out their genitals. Given the similar size of males and females, I asked Filippo how he sexed them. His answer was the opposite of what you'd expect. A monkey that has a "dangling genital thing" must be a female, he said, even at a young age. The penis and testicles of males are small and well-hidden underneath hair. We don't know the reason for this anatomical reversal. Sometimes females touch their own or another's pendulous clitoris, but it is unclear if the size of this organ gives them any extra pleasure.

While studying spider monkeys at the Chester Zoo, in the United Kingdom, Filippo regularly saw human parents explain to their children how well father monkeys take care of their babies. They'd point out a monkey with a genital appendage carrying an infant on her back and make up a story about the scene, as parents are wont to do. This would last until they read the zoo sign about the species's enlarged clitoris. Then they needed to figure a way to include the new information, if they chose to do so at all.

Chimpanzees and bonobos are easy to sex, especially a female in estrus. She will carry a soccer-ball-size pink signal on her behind that tells every male in the neighborhood that she's ready for action. The swollen perineal tissue and labia hide the clitoris, which is larger in bonobos than in both humans and chimpanzees. Female bonobos prefer face-to-face copulation and often invite males by lying on their backs, legs apart, a position that guarantees the stimulation of their frontal vulva. The evolution of male bonobos, however, must have been lagging behind. They favor the classic doggy position. This can cause comical confusion. If a male starts from the back, midway the female quickly turns around to get to her favorite missionary position. No wonder bonobo copulations are preceded by lots of gestures and vocalizations to negotiate positions. These Kama Sutra apes mate in every conceivable posture, including some that we're incapable of, such as hanging upside down by their feet.

I'm so used to ape genitals that they strike me as neither bizarre nor ugly, although I'd definitely call them cumbersome. Female apes with inflated genitals can't sit down normally; they awkwardly shift their weight between one hip and the other to avoid sitting on their swelling. The tissue is fragile. It bleeds on the slightest occasion but also heals remarkably quickly. We should be grateful that we've been spared these ornaments. If we'd had them, chairs would no doubt have been designed with a sizable hole in the middle.

The bonobo's clitoris begs attention because of the intense speculation surrounding its human equivalent. Initially, Sigmund Freud, the Austrian father of psychoanalysis, sidetracked us. He singlehandedly gave us a mythical source of pleasure known as the vaginal orgasm. Yes, this phenomenon was invented by someone who had limited anatomical knowledge and no vagina. Freud considered the vaginal orgasm superior, while dismissing clitoral pleasure as something for children.

Women who got pleasure from their clitoris without needing penetration were sadly stuck at an infantile stage, ripe for psychiatric treatment. Due to Freud's enormous influence, the clitoris was pooh-poohed as irrelevant. Medical textbooks depicted it as smaller than its actual size or erased it altogether.

Freud was wrong, however. The vagina, which connects the uterus to the vulva, isn't particularly sensitive. It serves as the birth canal and has a muscular wall that contains few nerve endings. It can't be a major source of pleasure. We've all heard of the G-spot, but so far no anatomist has been able to pinpoint its location. The clitoris, on the other hand, is easy to find. It is an erectile part of the vulva equipped with special cells adapted for sensory stimulation. Since the nerves of the clitoris reach into the vaginal wall—anatomists speak of a clitourethrovaginal complex—it's hard to know where exactly gratification arises. In contrast to the male orgasm, which is highly localized, the female's is diffused. Penetration may be an added source of pleasure but mostly thanks to friction against the clitoris, which is the jewel at the female orgasm's core.[8]

Freud's dismissal of the clitoris may have reflected a cultural worry that women would take control of their sexuality. Perhaps they'd tell men what to do or make them obsolete to their pleasure. An emphasis on penetration was a way of keeping women in line. As the American historian Thomas Laqueur put it:

> *The tale of the clitoris is a parable of culture, of how the body is forged into a shape valuable to civilization despite, not because of, itself. The language of biology gives this tale its rhetorical authority but does not describe a deeper reality in nerves and flesh.*[9]

Many feminists view the clitoris as empowering. The American science journalist Natalie Angier compared it to a well-tempered clavier playing godly Bach for any woman ready to listen.[10] Nevertheless, its function is not easily defined. Given that female orgasm isn't essential for conception, what good does it do? Some have argued that the clitoris is as useless as the male nipple and that women don't need it so long as they accept sex when it knocks on their door. That they achieve orgasm is a lucky by-product of evolution. The American philosopher Elisabeth Lloyd put it as follows:

> *Male and female both have the same anatomical structure for two months in the embryo stage of growth, before the differences set in. The female gets the orgasm because the male will later need it, just as the male gets the nipples because the female will later need them.*[11]

The biologist Stephen Jay Gould agreed with Lloyd that the clitoris hitchhiked on the evolution of the penis. He called the female orgasm a "glorious accident."[12] Gould, too, made a comparison with male nipples, which evolved as a by-product of the female's nursing capacity. All male primates, even the mighty gorilla, are equipped with nipples that they don't need and will never use. But most biologists, even though acknowledging the existence of vestigial traits, are skeptical when a natural feature is dismissed as nonadaptive. Our first impulse is that traits must exist for a reason. I feel the same about human body parts that hospitals routinely remove, such as the foreskin or the appendix. If these parts truly served as little purpose as the medical establishment believes, wouldn't evolution have removed them ages ago?

Regarding the appendix, the thinking has changed. This particular extension of the cecum has evolved more than thirty times in separate animal families, so can't be useless. It is thought to preserve gut flora that helps reboot the digestive tract after a severe case of dysentery. The appendix is now considered a functional part of the body.

I'd argue the same for the clitoris. First of all, we find one in every mammal. The mouse has a clitoris as well as the elephant. Second, it is an "expensive" organ. It is infinitely more involved and sensitive than the male nipple. It is a marvel of evolutionary engineering. Lloyd and Gould didn't know this when they argued their case, but the clitoris is on a par with the penis regarding the thousands of nerve endings that pick up its signals. It is fed by remarkably thick nerves, indicating its relevance to both body and mind. Since it carries an even higher density of sensory cells than the penis, it doesn't seem accidental at all.[13]

The clitoris likely evolved to turn sex into a pleasant, addictive affair. The assumption here is that of an enterprising female sexuality, one that seeks until it finds what it likes. This would explain why the largest clitorises are found in species marked by multipurpose eroticism. Apart from ourselves, this holds for dolphins and bonobos, both of which frequently engage in genital stimulation, sexual petting, or outright intercourse for the sake of bonding and peaceful coexistence. I don't think it's by chance that the dolphin clitoris is the largest one known in nature.[14] Nor do I consider it a coincidence that bonobos have such a prominent clitoris, which in young females sticks out frontally like a little pinky finger. Later in life, it gets embedded in the surrounding swelling tissue and is harder to spot, but it still doubles in size at times of arousal. From limp and soft, it becomes rigid and stiff. Since both its glans and its shaft harden, the bonobo's clitoris responds to stimulation like a penile erection. During intercourse with a male, a female bonobo often reaches down with a hand to stimulate either her partner's testicles or herself.

Laboratory experiments with monkeys show that we're not the only species in which the female heart beats faster when sexual intercourse reaches its climax. Monkeys also show uterine contractions at this moment, thus meeting Masters and Johnson's criterion for orgasm. No one has tried the same experiments with bonobos or dolphins, but they, too, would doubtless pass the test.[15]

Anyone who witnesses two female bonobos in the midst of intense GG rubbing will agree that it looks extremely enjoyable. The females bare their teeth in a grin and utter piercing squeals as they frantically rub their clitorises together while staring into each other's faces. Detailed video analyses by Sue Savage-Rumbaugh at the Yerkes Primate Center have shown how vital these exchanges are. Sexual contacts are mutually initiated and collaborative. When a bonobo male and female copulate, the male's speed of thrusting goes up or down in response to the female's facial expressions and vocalizations. He may stop thrusting altogether if she avoids eye contact or signals boredom by yawning or grooming herself. In chimpanzees, in contrast, the male dictates the position, and eye contact occurs only when the female looks back over her shoulder.[16]

The biggest giveaway about pleasure is that bonobo females regularly masturbate. Lying on their back, they rhythmically move a finger or toe through their vulva while staring into the distance. This leisurely activity, which lasts a lot longer than the typical copulation, would make no sense unless they were getting something out of it.

ON A SUNNY DAY, my chimpanzees are delighted to see me. Or rather, my sunglasses. They hurry over to pull weird faces at seeing their reflections. They gesture for me to remove these little mirrors and hold them closer. Apes belong to the handful of species that recognize themselves in a mirror. They open their mouths to stare into them, picking with a finger at their teeth. Females will turn around to inspect their behind, especially if it is swollen. It is a hugely important part of their anatomy that they normally don't get to see. Males never turn around. They have no interest in their own behinds.

A bonobo or chimpanzee female in estrus gives the impression that she knows exactly what flag she is carrying. She walks around while sticking her genitals triumphantly into the air by arching her back. She bends over a little too often to pick up items. That's what you get with self-aware animals: they realize how they come across to others. Conversely, a female may try to hide her assets in the presence of males whom she doesn't want to seduce. Wild female chimps, for example, avoid mating with senior males with whom they grew up. They retreat screaming from these potential fathers while being perfectly accepting of younger males' courtship.[17]

In our chimpanzee colony, young Missy developed such an aversion to Socko. Each time her genitals were swollen, she'd do her so-called "crabwalk." She'd walk all hunched up, sometimes sideways, making her swelling almost disappear between her legs, which is hard to do. At first, we thought she might be sick or had perhaps broken a limb. But we soon discovered that she walked in this odd manner only when two conditions were met: when she was swollen, and when Socko was around. Old enough to be her father, Socko was the group's alpha male. We speculated that Missy wanted to avoid his attentions, which she often retreated from. If this tactic failed, her mother, May, helped out. When copulation was about to take place, May would rush over with distress yelps and pry the two of them apart with her hands. May herself didn't mind mating with Socko, but her daughter was a different matter. May backed Missy's repulsion.

Every female ape's genital swelling has a different color, shape, and size. We learned the significance of this while we were exploring individual recognition. Instead of focusing on faces, as so many previous studies had done, we decided to include behinds. We first trained chimpanzees to select matching pictures of flowers, birds, and so on, on a touchscreen. Once they were good at this, we showed them a picture of chimpanzee buttocks followed by two portraits. Only one portrait came from the same individual as the rear view. Could they apply the matching rule to these pictures as well?

The chimps had no trouble linking the correct face with the behind. It is telling that they only did so with chimps they personally knew. That they failed with pictures of strangers shows that their choice was not based on something in the pictures themselves, such as color, size, or background. Rather, it reflected intimate knowledge of their fellow apes. We concluded that apes possess a whole-body image of familiar individuals. They know them so well that they can connect one part of another's body with any other part. We do the same, such as when we locate friends in a crowd even if we only see them from the back.[18]

When we published our findings under the title "Faces and Behinds," everyone thought it was funny that apes could do this. We were awarded an Ig Nobel Prize—a parody of the Nobel Prize that honors research that "first makes people laugh, and then think." A follow-up study by the Dutch primatologist Mariska Kret brought us full circle to Morris's claims about the eroticized human face. Kret used touchscreens to compare the recognition of faces and behinds in both humans and chimpanzees. The apes were better at recognizing the behinds of their species than human subjects were with human behinds. Kret thinks this is because in the course of evolution, our ancestors attached less and less importance to their behinds and shifted the focus to the face.[19]

The eye-catching genitals of apes are the product of sexual selection. This type of selection differs from natural selection. Natural selection favors traits that assist survival, such as camouflage colors and escape tactics, not flamboyant signals that can be seen from a mile away. If survival were the issue, the unwieldy genital swellings of chimps and bonobos would never have come into existence. They make it hard to climb around and sit down. They serve only to call attention. This is no minor issue, though, in relation to finding mating partners. This is why Charles Darwin proposed a second selection mechanism.

Sexual selection favors traits that do nothing for survival but appeal to potential mates. Good examples are extravagant male ornaments and behaviors, such as the peacock's tail, the decorated nests of male bowerbirds, and the elaborate antlers of stags. These traits handicap their owners while making them more visible. The only reason they stay in the gene pool is that females like them. More than that, females *insist* on them. A male whose colorful tail is not up to par or who can't perform the right song and dance can forget about gaining her attentions. Female bowerbirds are comparison shoppers, who inspect many nests in their area before settling on a male worth mating with. Most beauty in nature exists thanks to female taste.[20]

Whereas in most animals the male is splendid and the female drab and camouflaged, our little hominid trio—humans, chimpanzees, and bonobos—seems to have turned things around. We have shifted beautification from the male to the female. It is the female who is ornamented and judged by it. Sexual selection can of course go both ways, but for it to be reversed, males need to have outspoken preferences. Male apes are indeed obsessed with female behinds. It is not unusual to see five or more of them closely follow a swollen female around. It's a huge magnet. Not surprisingly, in the above touchscreen experiments, the greatest connoisseurs of behinds were males, not females.

Men, too, are obsessed with the body shape, buttocks, breasts, and face of the other sex. These features have the power to take their breath away. This is why many more establishments offer men a chance to watch unclad female bodies than vice versa. Conversely, women are conscious of their bodies and compare their appearance with that of other members of their sex far more than men do.[21] In modern society, women spend so much time and money on beautification that a multibillion-dollar fashion, cosmetics, and plastic surgery industry caters to their needs—or, as some would say, exploits their insecurities.

Even though women differ from apes in that they lack body signals of fertility, they make up for this through the clothes they wear. American university students were photographed at different points in their menstrual cycle as determined by self-report and urine tests. Judges of both genders were then asked to pick out the photos in which these young women seemed to "try to look more attractive." It turned out that efforts to enhance their appeal changed with the cycle. Around their ovulation peak, the pictured women wore fancier, more fashionable clothing and revealed more skin. An Austrian study found a similar tendency. Investigators concluded that fertility unconsciously pushes women to boost their appearance and ornamentation.[22]

This raises the question of whether female apes, too, embellish themselves. I know of no systematic studies, but even a cursory glance at the literature reveals how common self-adornment is. I myself have often seen chimpanzees pick up unusual objects, ranging from colorful feathers to a dead mouse, place them on their head, and walk around thus decorated for the remainder of the day. They also often drape vines and branches around themselves or put them on their backs. The majority of these chimps are females. A pioneer of animal cognition, the German psychologist Wolfgang Köhler, described how his chimps turned "impishly self-important or audacious" after dressing themselves with branches, ropes, or chains.[23] Robert Yerkes, too, relates that adolescent female chimps would crush colorful fruits, such as oranges or mangoes, and adorn themselves by placing them on their shoulders. This was not just a visual signal but also an aromatic one.[24]

At a chimpanzee sanctuary in Zambia, this sort of behavior developed into a group-wide fashion. One female stuck a straw of grass into her ear, letting it hang out like jewelry while she was walking around and grooming others. Over time other chimps followed her lead by adopting the same grass-in-ear "look." Of the hundreds of recorded instances, 90 percent involved females.[25]

The level of self-awareness in dress-up games is striking. At a facility with sign-language-trained chimpanzees, two young females were undoubtedly inspired by humans when they put on glasses and applied lip gloss while inspecting themselves in a mirror.[26] The German scientists Jürgen Lethmate and Gerti Dücker described how Suma, an orangutan at the Osnabrück Zoo, spontaneously responded to a mirror placed near her cage:

> *She gathered salad and cabbage leaves, shook each leaf, and piled them up. Eventually, she placed one leaf on her head and walked straight to the mirror with it. She sat down directly in front of it, contemplated her headcover in the mirror, straightened it a bit with her hand, squashed it with a fist, then put the leaf on her forehead and began to bob up and down. Later, Suma arrived holding a salad leaf in her hand at the bars to lay it on her head once she could see herself in the mirror.*[27]

Apes raised in human families (a practice that fortunately has disappeared) carry blankets, even on sweltering days, and decorate themselves with hats, saucepans, paper bags, or other kitchen items.[28] I realize that all the above examples could reflect human influence, but we also have a few observations from the field. Sometimes the ornaments aren't so pretty, such as a dead snake or the intestines of a recently killed forest antelope. A wild bonobo female was seen wearing the latter as a necklace. Similarly, a young female chimpanzee at the Mahale Mountains, in Tanzania, put a knot in a strip of monkey skin before she threw it around her neck and walked around with it.[29]

It's not that males never augment their presence, but they do so for a different reason. At one field site, for example, a male chimpanzee stole empty kerosene cans from camp and noisily banged them together. By scaring the bejesus out of everyone, he managed to rise in status. Male apes in the wild may wield a big stick or branch during bluff displays. At zoos, they often use empty buckets to drum on or kick around. Their selection of accessories has less to do with sex appeal than with status and intimidation.

Awareness of one's appearance and an interest in its embellishment seems mostly a feminine trait.

JUVENILE FEMALE CHIMPANZEES stroll around with the babies of others and play with their peers, but most adults don't pay them any heed. All this changes, however, once their first little genital swelling develops at the age of nine or ten. Male eyes then begin to follow them around. The pink balloon on their butt grows larger with every consecutive cycle. At the same time, they become sexually active. At first, they have trouble seducing adult males and are successful only with adolescent ones. Their insatiable sexual curiosity exhausts any young male who shows interest. It is not unusual to see a young female tweak a male's penis with her fingers once it has begun to sag after a day of unabated demands.

The larger a young female's swelling grows, the more she begins to intrigue the grown males. The female quickly learns that this gives her a leg up in the world. In the 1930s, Yerkes conducted experiments on what he called "conjugal" relations in chimpanzees (a misnomer, because this species lacks stable bonds between the sexes). After dropping a peanut between a male and a female ape, Yerkes noted that swollen females enjoyed privileges that females without this bartering tool didn't have. A female chimp with a genital swelling had no trouble claiming the prize. Outside her swelling phase, however, the male was in control of the peanut supply. Yerkes concluded that signs of fertility allow females to cancel male dominance.[30]

Publication of this study led to an amusing counterpoint by the American poet Ruth Herschberger, who conducted an imaginary interview with Josie, Yerkes's main subject. Josie the chimpanzee disagreed that the colossal male with whom she had been paired was "naturally dominant." In the course of many tests, she had gathered about as many peanuts as he had. Josie speculated that her success was not due to feminine wiles but that she simply became braver and more assertive when she was fertile. She was particularly offended by the word *prostitution* that Yerkes had dropped in one of his descriptions: "It's this prostitution angle that makes me the maddest!"[31]

The outcome of Yerkes's experiment was no aberration, however. Status changes associated with the female cycle occur in nature as well. Referring to wild chimpanzees, Goodall noted that "the swollen state is very definitely associated with a variety of privileges for the female concerned." She offered striking examples, such as old Flo, who normally never competed over the bananas that were provided at camp. When swollen, however, she'd push in among the big males to claim her share.[32]

Whenever chimps capture prey, male hunters share the meat preferentially with swollen females. When such females are around, male chimps hunt more avidly because of the sexual opportunities. A low-ranking male who captures a monkey becomes a magnet for the opposite sex, which offers him a chance to mate for meat until he's found out by someone higher ranking than himself. At Bossou, in Guinea, males have few hunting opportunities but raid the papaya plantations of surrounding farms. This perilous undertaking allows them to share delicious fruits with fertile females.[33]

Similar tradeoffs occur in bonobos, but mostly with immature females. I once photographed an adolescent female grinning and squealing during a face-to-face copulation while her partner held two oranges, one in each hand. The female had presented herself to him as soon as she saw the goodies. She walked off with one of the fruits. The reason a young female bonobo's self-confidence fluctuates with the size of her genital swelling is that she doesn't yet dominate any adult males. It may be a vestige of a past in which female bonobos still used sex, as female chimpanzees do, to barter for favors. After overthrowing male rule, this tactic must have lost its appeal. Most adult female bonobos don't beg for male favors: they just claim what they want.

The growing sex appeal of young female apes finds a parallel in our species when a teenage girl's bosom begins to expand. She, too, becomes a magnet for male attention and learns the power of cleavage. She goes through emotional upheavals and insecurities similar to those of an adolescent female ape. Her changing body prompts a complex interplay between power, sex, and rivalry. On the one hand, her appearance may give her the sort of clout with males that she's never enjoyed before. On the other, it produces unwanted attention and risks. Like the chimpanzee Missy, she may want to hide her body from leering males. A further complication is the rising jealousy of other girls and women. All this is brought about by the blossoming of unmistakable feminine body signals. The main difference between humans and apes in this context is that most of our signals are concealed, since we don't publicly flash our genitals.

This is not entirely true, though. Manspreading may be an unconscious genital display without showing the goods, but actual exposure of male genitals is not unheard of. We have learned from the #MeToo movement how often men send uninvited pictures of their member or pull it out in the presence of unsuspecting women. As in other primates, this kind of exhibitionism is both a solicitation and a form of bullying and intimidation. Women, too, sometimes publicly show their breasts or genitals or at least hint at them. But most of all, our face has become the key area of signaling.

Our face contains a plethora of gendered signals, which is why we are so fast and accurate at classifying faces by gender. We recognize men by their more robust jaws, which give their face a square shape compared to the more oval face of women. Moreover, women's eyes are relatively large, and they also have larger pupils. Long eyelashes further accentuate women's eyes. Women's facial features (eyes, lips) also contrast more with the surrounding skin, which is thinner and softer than a man's.[34]

As if these natural differences weren't enough, we magnify them in ways that turn our faces into major gender signposts. Men grow beards, or they shave off all the hair on top of their heads, sometimes both. Men without beards may still like the stubbled look, which is seen as rugged and badass. Women, by contrast, grow their hair long while meticulously removing all facial hair. Many of these trends are cultural dictates, and my description here focuses on the West, where even the slightest peach fuzz on a woman's upper lip has to go. Women also epilate and

The human face is a signpost of gender. Even after removal of cultural markers, such as hairstyle and makeup, we instantly recognize facial gender. It is expressed in the overall shape of the face (square vs. oval) as well as the relative size of eyes and lips.

wax their eyebrows to make them unlike men's bushy ones. Eyes may be emphasized with false eyelashes and mascara, thus mimicking the doe-eyed look of infants. The habit of women to paint their lips red to make them look fleshier is thousands of years old. It probably goes back to the ancient Egyptians, who used red ochre, carmine, wax, or fat. When lipstick became too expensive during World War II, women stained their lips with beet juice.

Owing to all these cultural modifications of the facial appearance, an individual's gender is typically broadcast far and wide. It's all part of an evolutionary history in which an upright gait required that sexual signals be relocated on the body. The signals traveled from rear to front and from bottom to top where they could get the attention they deserve.

7

THE MATING GAME

The Myth of the Demure Female

Whenever people talk about self-esteem, the first image that flashes through my head is that of the self-assured old boss of a large rhesus monkey troop named Mr. Spickles.

For a decade, I worked with macaques at the Henry Vilas Zoo, in Madison, Wisconsin. Spickles was a fully self-actualized kind of guy, whose name came from the red freckles covering his face. He moved around the rocky outdoor enclosure in a dignified manner, surrounded by females eager to groom him. Reclining with his legs spread wide, Spickles would show off his scarlet scrotum and close his eyes during the lice removal diligently performed on him. He appeared twice the size of any female, but most of this was hair. He always walked with his tail proudly in the air, something no other male dared to do, at least not in his presence. At the same time, however, his position was up to the females. Orange, the troop's alpha female, fiercely supported him. A macaque society is essentially a female kinship network run by the top matriline.

The reason I mentioned "self-actualization" is that about a century ago, these monkeys were studied at the same little zoo by Abraham

Maslow, the psychologist who gave us the description of the hierarchy of needs. Only when all your basic needs (for safety, belonging,
prestige) are fulfilled, he proposed, will you be able to realize your full
potential. Few people know that this staple of business seminars was
inspired by Maslow's observations of the top monkey's cocky, confident
air and the "slinking cowardice," as he called it, of individuals near the
bottom of the social ladder. Turning his attention to us, Maslow translated monkey self-confidence into human self-esteem. This blend of
self-assessment and navel-gazing struck a chord with American culture
that lasts to this day.[1]

The paradox that an individual could be dominant yet dependent on
others probably never crossed Maslow's mind. Like most psychologists,
he thought in terms of individual traits and personality types. Dominance is a *social* phenomenon, however: it resides in the relationship,
not in the individual. One cannot lead those who refuse to follow. So
instead of viewing Spickles as imposing his dominance on the rest, it's
better to consider him the accepted dominant. He gained the respect
and support of everyone, including Orange. And the fascinating thing
was that even while she kept him in the saddle, her sexual interests were
a different matter altogether. During the mating season, she was drawn
to younger males.

Native to temperate regions of Southeast Asia, rhesus monkeys mate
in the fall so that infants are born in the spring. When females become
fertile, life in a troop changes dramatically. The females seek out males
with whom to mate while competition among the males intensifies.
Males often interrupt the matings of others below them in the rank order.
Throughout one mating season, a certain triangle held my fascination:
Spickles, Orange, and Dandy. The first two were well-established characters. Named after her bright hair color, Orange was the most-watched
individual in the troop. Whenever she walked around, other females
reacted by withdrawing their lips from their teeth and grinning from
ear to ear. Macaques grin in order to appease high-status individuals.
Grins send a message of unequivocal submission, removing any need

for the dominant macaque to enforce her position. Orange received far more grins than Spickles, but since she herself occasionally grinned at him (and he never at her), he formally ranked above her.[2]

Dandy was a handsome, vigorous male, less than half Spickles's age. He could race around the large outdoor enclosure and climb upside down along its mesh roof with a speed and agility that no one could match. Least of all the alpha male Spickles, who was stiff, slow, and quickly out of breath. Spickles had trouble handling Dandy, who sometimes provoked him by jumping right in front of him or by standing his ground if Spickles threatened him. Every time such a scene took place, Orange would calmly walk up to them and take a position right next to Spickles. She didn't need to do more than stand by his side, because Dandy knew he could never win this confrontation. All the females would back Orange. Going against the alpha female is not an option in the strict rhesus hierarchy.

During the mating season, however, Orange specifically sought out Dandy for mating. Spickles tried to prevent this, chasing off the younger male (without ever catching him), but Orange would simply return to Dandy to hang out with him. The two of them would huddle together for days, while Orange occasionally pushed Dandy to activate him. She'd present her behind so that he could mount her. The longer these consortships lasted, the more Spickles resigned himself to them. Sometimes he voluntarily left the scene by going indoors for a while, thus giving the two lovers room to copulate without worries. My diaries from this period show that, as a young scientist, I was baffled. Why did Spickles remove himself? My speculations ranged from him attempting to "save face" to him being unable to stand the sight of their coupling. Perhaps he was engaging in stress management. By the end of the season, Spickles had lost 20 percent of his weight.

We often look at the social life of monkeys as simple compared to that of the apes, but I have learned never to underestimate monkey sophistication. In this particular love triangle, Orange carefully balanced two preferences: one concerned political leadership, and the other sexual desire. She never confused the two. Twice I saw Dandy take advantage of his proximity to Orange by challenging Spickles. Both times Orange immediately corrected her young lover. For good measure, she also attacked Dandy's mother, as if to make the point that his whole family had better know their place.

ALTHOUGH MY TEAM saw Spickles mate far more often than any other male, he didn't father more offspring. We know this because, for eight years, this troop was part of one of the first paternity studies in primatology. Traditionally, primatologists looked at alpha males as those who succeeded in spreading their genes. But in making this point, we relied entirely on observed sexual activity. The more we saw a male mate, the more offspring he sired, we thought. This assumption proved flawed. While alpha males have no qualms about mounting females in the open, other males often get busy out of sight and at night.

At that time DNA technology was not yet available, but scientists at our primate center compared the blood groups of newborns with those of potential fathers. We found a rough correlation between a male's rank and the number of offspring fathered. Alpha males did do better than average, but they weren't nearly as successful as we had predicted. Up-and-coming males—such as Dandy—sometimes sired more progeny.[3]

Position in the male hierarchy is only one factor in the mating game. The other one is female preference. This factor was long overlooked, partly because female choice is harder to see than male bluster. Few females can act with the impunity of Orange, because they run a risk if their sexual preferences don't match the male hierarchy. Trysts with males down the ladder require evasive tactics. "Sneak copulations," as they are known, take place behind the bushes or while the boss is asleep. Primate groups brim with illicit sexual activity. I have often watched the scenario play out among chimpanzees.

A few meters from a male, a female will casually lie down in the grass, her genital swelling pointing at him. As if nothing is the matter, she'll look over her shoulder while he nervously checks around to see where the dominant males are hanging out. Being close to a female in this state is risky in itself. The chosen male slowly gets up and strolls away in a particular direction, occasionally stopping to look around furtively. A couple of minutes later, the female will walk off in a different direction. She knows exactly where the male went, and via a detour, she'll meet up with him. They will have a quickie at a hidden spot, then go their separate ways. Except for a few curious youngsters of their species and the human observer, no one is the wiser. Their deception is supremely cooperative, including the suppression of sound. Chimpanzee females typically vocalize at the climax of intercourse but never during a secret rendezvous.[4]

The second reason we have underestimated the role of female choice is cultural. Both in biological science and in society at large, the female sex, whether animal or human, was depicted as passive and coy by nature. More than that: females were *expected* to be passive and coy. Exceptions were minimized or overlooked. Who got to mate and who didn't was seen as a male decision. Females might play hard to get, allowing them to select the best male from among several suitors, but female sexual initiative wasn't part of the era's biological theories.

That we thought this way for so long is lamentable given that Darwin had already proposed a broader view. It was ignored and suppressed for over a century. Darwin may have shared the dim opinions about females popular in his time and place, especially with respect to their intellectual powers, but he was miles ahead when it came to estimating their role in evolution. He was the first biologist to stress female agency. While everyone else was seeing females as mere vessels of male reproduction, Darwin developed sexual selection theory, according to which we owe nature's brilliant colors and pleasing songs to female preferences for male behavior, ornamentation, and weaponry. By mating with the best-endowed males, females steer evolution. Darwin's contemporaries ridiculed this idea, which assigned females a crucial role. The English botanist St. George Mivart was sure that "such is the instability of a vicious feminine caprice, that no constancy of coloration could be produced by its selective action." Given that in those days *vicious* meant "wicked," Mivart was essentially accusing Darwin of advancing an immoral proposal.[5]

In addition to their lack of confidence in females, critics felt that "brutes" (animals) lacked freedom of choice. It was patently absurd to think that female birds, or any other animals, could decide anything. This view was amplified by the previous century's low opinion of animal intelligence in general. Animals were depicted as machines, driven by instinct and simple learning. Labs full of lever-pressing rats and stimulus-pecking pigeons proved just how dumb they were. It was ridiculous to expect them to make fine-grained choices about anything other than perhaps what to eat.

Anthropologists didn't help either. They saw women as mere pawns in men's games. The dominant theory held that daughters and sisters were the property of men. They were exchanged as "supreme gifts" to cement alliances between patriarchal groups. We still live with a symbolic leftover of this attitude during weddings, when the bride is "given away" by her father to the new husband.[6]

The view that the mating game is played among males, and that females are its passive objects, remains immensely popular despite lack of evidence. The first scientific holes in it were blown by work on the same animals that had inspired Darwin: birds. In the 1970s, scientists wanted to control a population of red-winged blackbirds. Having vasectomized some of the males, they expected to find sterile clutches of eggs. But when they incubated eggs from these males' nests, they were shocked to discover how many hatched.[7] Who could have fertilized them? Had intact neighboring males perhaps forced themselves on those poor females?

So entrenched was the belief in feminine passivity at the time that the researchers thought sex outside a pair could only be involuntary.

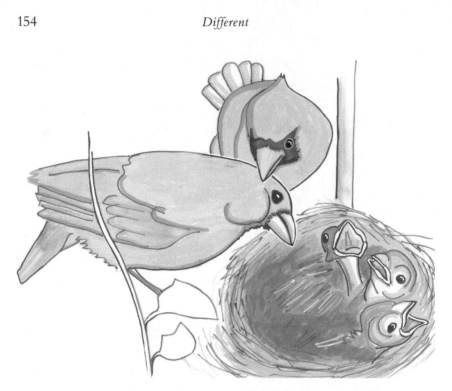

Songbirds, such as these Northern Cardinals, are routinely held up as models of monogamy. DNA testing, however, has revealed that clutches often come from multiple fathers. Females are as sexually adventurous as the males.

The more birds scientists studied, however, the more they discovered clutches that had been fathered by multiple males. Moreover, the idea that females were victimized by marauding invaders fell apart. When birds were followed with radio tracking, the truth emerged. Working with hooded warblers, the Canadian ornithologist Bridget Stutchbury saw females actively pursue outsiders. They would make forays away from their nest while loudly calling, as if telling potential mates, *Hey, I'm over here!*[8]

These observations were all the more impactful since bird monogamy has traditionally been held up as an inspiration for humankind. A century ago an English reverend held up the pair-bond of the common hedge sparrow, or dunnock, as the perfect example. We'd all be better off if we acted more like those sweet little birds, he told his flock. Even as an amateur naturalist, however, the reverend didn't have a realistic picture. He didn't know what we have since learned from the world's expert on dunnocks, Nick Davies of Cambridge University. Having documented the many ménages-à-trois and dalliances of these birds, he has made it clear that it's not just the males. Female dunnocks take an active part in their racy sex lives. Davies surmised that if people had taken the English reverend's advice, "there would have been chaos in the parish."[9]

The female sex drive of birds is so underrated that its recognition can yield lots of money. Pigeon racing, a popular sport both in Europe and in China, is conducted over long distances, such as from Barcelona to London or from Shanghai to Beijing. The first bird to get home wins a huge prize. Female sexual desire came up in an interview with the Belgian owner of New Kim, a champion pigeon for whom a Chinese billionaire paid nearly $2 million. The proud pigeon fancier explained that racers traditionally apply a "widowhood" technique to male birds. A few days prior to a race, they separate the cock from his mate to enhance his homing motivation. New Kim was a female, but her owner had discovered that the same technique worked with her as well. He kept her from mating with her cock for several days while allowing her to see him. It was the only way, he said, to make her fly faster than the rest. She'd be eager to return home to "party" with her mate.[10]

Recognition of the existence of a sex drive in female birds set the stage for *Darwinian feminism,* as the American biologist Patricia Gowaty dubbed it in 1997. This label may sound like an oxymoron because many feminists consider humans to be far removed from the birds and the bees. They don't see evolutionary science and its emphasis on genetics as particularly friendly to their cause. But for biological scientists, including the feminists among us, feminism can't escape a connection with biology. After all, there wouldn't be any need for feminism if we didn't have two sexes to begin with. And why do we have two sexes? Because sexual reproduction works better than its alternative, which is cloning. As a cloning species, we'd be free of gender inequality because we'd all look the same and reproduce the same way, but we'd pay an enormous price.

For good reason, sexual reproduction evolved more than a billion years ago in both plants and animals. It's so widespread that most of what we know about it didn't come from our own species. The laws of inheritance, for example, were discovered by a Silesian monk growing peas. Having two parents contribute to reproduction shuffles the gene deck with every new generation, allowing descendants to carry novel genetic combinations and to be ready to meet a changing environment and new diseases. It makes us genetically flexible.

Without sexual reproduction, we'd be equal but unprolific.

DARWINIAN FEMINISM SEEKS a more inclusive account of how the interplay between the sexes drives evolution. Why this topic deserves attention is not always understood, however. In the 1990s, Gowaty participated in a seminar for a women's studies program in Kentucky, where she compared the contributions of males and females to reproduction. Afterward an angry critic confronted her, insisting that evolutionary arguments were beside the point and that everything Gowaty had said could be explained by men's fear of women's sexuality. This take was not too far-fetched given Freud's contempt for the clitoris, the slow recognition of female sexuality in birds, and efforts to erase those "exhausting" bonobos from the story of human evolution. Society doesn't welcome female sexuality, and men of science have systematically tried to lock the female libido in a box and throw away the key.[11]

But Gowaty and her critic could both be right. Most people think at the level of everyday psychology, which is quite different from the evolutionary approach. To understand evolution, it is essential to step back from what drives behavior in the here and now. Instead of considering the motivations, ideology, upbringing, experience, culture, hormones,

Above: Primates devote much time and attention to grooming, which is the glue of their society. An adolescent male bonobo grooms an adult female.

Left: Female apes have a tendency to embellish their appearance. A young bonobo self-adorns by draping banana leaves around her shoulders.

Left: Juvenile bonobos tongue-kiss during a playful erotic encounter.

Right: A bonobo female advertises fertility by sporting a large balloon-like swelling on her behind, which is a water-filled edema of the external genitalia. This conspicuous pink signal attracts males.

An adult male bonobo (left) manually stimulates a younger male who presents his erect penis.

A large gathering of wild females (and their young) from three separate bonobo communities. Peaceful inter-group mingling is common at the Wamba field site in the Democratic Republic of the Congo. (Photograph courtesy of Takumasa Yokoyama and Takeshi Furuichi.)

Claudine André with Princess Mimi, the reigning alpha female of the first bonobo group at Lola ya Bonobo, near Kinshasa. As the founder of the world's only bonobo sanctuary, André has successfully rescued many orphans and reintroduced them to the wild once they have grown up. (Photograph courtesy of Christine l'Hauthuille [Comité OKA-ABE].)

Since bonobos often mate belly to belly, facial communication plays a greater role than in other species. Here an adult male is positioned on top of a female, but the reverse also occurs.

Bonobo mothers nurse their babies for four to five years.

Wild female baboons often have male friends who protect them. The female on the right has moved out of the way of an aggressive new young adult male on the left. She clings to the back of her friend, who stares down her harasser.

Female primates are fascinated by newborns. A mother stumptail macaque (the central female with the prominent nipples) is surrounded by other females, young and old, who grunt in chorus every time the baby does something surprising (such as sticking his foot in his mouth), as if commenting on the marvel of new life.

Friendships between female chimpanzees, which may last a lifetime, are serviced by long grooming sessions. The female on the right is Kuif, Mama's best friend and the adoptive mother of Roosje.

Left: I am holding baby chimpanzee Roosje before her adoption by Kuif at the Burgers' Zoo. (Photograph from 1979, courtesy of Desmond Morris.)

Right: Kuif got so good at bottle-feeding Roosje that she later raised her own offspring the same way.

Left: Luit was one of the finest alpha males I have known. His tragic end illustrates the fierce status competition among male chimpanzees.

Right: Even in a male-dominated society, the alpha female can be a powerful leader. With her commanding personality, Mama enjoyed enormous authority in the chimpanzee community.

Sexual dimorphism is the difference in size and appearance between the sexes. An adult male chimpanzee (left) is seen next to an adult female (right). Whereas males are hairier and heavier than females, the average size dimorphism in chimpanzees is only slightly greater than in humans.

Top: A peace offer high up in a tree between two adult male chimpanzees who had been in a fight. One male holds out an open hand to his rival. Immediately after, they kiss and embrace, then climb down together to complete their reconciliation with a grooming session on the ground.

Left: A mother chimpanzee holds her daughter still while intently grooming her head.

Top: A weaning compromise between a mother chimpanzee and her four-year-old son. After repeated nursing conflicts, the son is permitted to suck on a part of the mother's body other than the nipple. This phase will last only a few weeks before the son loses interest.

Left: Mother is always there to help. A female reaches out to her son who has trouble descending from a tree.

Donna is a gender-nonconforming chimpanzee of the female sex with a masculine body and habits. She often displays side by side with adult males, with all her hair on end. Donna is nonaggressive, though, and socially well integrated. (Photograph courtesy of Victoria Horner.)

Left: Male chimpanzees don't tolerate the sexual escapades of young males. An adult male punishes a juvenile, who stayed too close to a swollen female. He bites the transgressor's foot while swinging him around.

Right: Apes show good intentions by laughing during play. Their facial expression resembles the human laugh as do the sounds they utter.

When, in Kibale National Park in Uganda, young chimpanzees lost their mothers to respiratory disease, some were adopted by their big brothers. On the right is Holland, a seven-year-old (prepubertal) male who was taken care of and protected by seventeen-year-old (young adult) Buckner, on the left. (Photograph courtesy of Kevin Lee and John Mitani.)

Adolescent female apes practice mothering skills on the young of others. Amber (middle) carries Mama's daughter, while her best friend (left) is in charge of a juvenile male (right). Both babysitters are still too young to have their own offspring.

feelings, and so on that go into our decision making, evolutionary biologists think in terms of millions of years. They think long term and try to peek behind the *veil of evolution* to consider behavior's genetic background. How does it promote survival and reproduction? They don't care about actors' motivations or whether they even know about these long-term benefits.[12]

A pertinent example is sex. We engage in it for two reasons, only one of which drives us in the moment. The first is sexual attraction and desire. Intense physical changes engorge and lubricate us in preparation for the acrobatics that we call lovemaking. Our goal is to satisfy our urges, experience pleasure, resolve social tensions, express tender feelings, and so on. These are the lusty motives that we all know and understand.

Our second reason for having sex is behind the veil. It is why sex exists and why we share its curious mechanics of insertion and thrusting with so many other species. Sex is the way we arrange a meeting between sperm and egg to make a zygote. This meeting isn't part of our motivation. Except for the times when we deliberately try to conceive, reproduction may not be on our minds at all during sex. That's why someone had to invent the morning-after pill.

For animals, the veil of evolution is thicker still—it is opaque. We have no evidence that any species other than our own knows that sex leads to progeny. While we can't entirely rule it out, the interval between the two is likely too long for other species to make the connection. This means that reproduction isn't what drives sex. Even though we call animals' sexual activity "breeding," this is only how *we* see it, not the animals themselves. For them, sex is just sex. Mothers obviously know their offspring since they gave birth to them and nursed them, but this is not based on any knowledge about fertilization. Fathers know even less.

Given this limited understanding, it's annoying that nature documentaries act as if animals do know. Over footage showing two zebra stallions rising to kick and bite each other, a narrator intones with authority,

Two zebra stallions engage in a fierce battle while the mares keep grazing. These fights are about mating and only indirectly about breeding. Zebras are unaware of the connection between sex and reproduction.

"These males are in combat over who will fertilize the mares." Zebra stallions, however, are unaware of sperm, eggs, genes, or how pregnancies come about. They fight over who will mount a mare. Period. The question of who will sire offspring is none of their concern. Only we, biologists, look behind the veil and think in terms of which male will pass on his genes.

At some unknown point in time, probably thousands of years ago, our ancestors began to realize that pregnancy requires sex. But how exactly the two were linked remained obscure for most of our history and prehistory.

After much hesitation and with many guilty feelings, the Dutch scientist Antonie van Leeuwenhoek placed some of his own ejaculate under his new invention, the microscope. He discovered thousands of wriggling "animalcules" inside. This happened in 1677, which goes to show how recent our current knowledge is. Darwin didn't know about genes or how those of both parents interact. He assumed that eggs and sperm received information from all over the body, which was then blended and handed down to the next generation. Modern genetics replaced pangenesis and other such theories only after the work of Gregor Mendel, the friar in the pea garden, in 1900.[13]

Our fellow primates aren't oblivious to every aspect of reproduction, though. They do have firsthand experience with pregnancy, parturition, and nursing. Older females, in particular, probably know all the stages that pregnant females go through. But even individuals without direct experience may know more than we think. I got my first inkling when I saw a young male capuchin monkey, Vincent, walk up to his best female friend, Bias, and deliberately place an ear on her belly. He kept it there for perhaps ten seconds. In the following days, I saw him do this several times. At the time, I didn't know that Bias was pregnant (which is hard to detect in these monkeys), but a few weeks later she had a tiny newborn on her shoulder. It's unlikely that Vincent had recognized her pregnancy by smell (like us, monkeys rely mostly on vision), but he could have felt the fetus moving around while huddling with his friend. I guess he wanted to hear the fetus's heartbeat.

In apes, I have noticed a similar interest in expectant females. Since apes also have midwifery, they seem aware of what is coming when one of them is pregnant. This still doesn't amount to grasping how reproduction works, though. While discussing evolutionary explanations of primate behavior, it is always crucial to distinguish between what *they* know versus what *we* know. And even for our species, which does realize that sex equals babies, most of our behavioral origins remain cloaked by the evolutionary veil.

SAINT SIMEON STYLITES, a fifth-century ascetic, is said to have lived for thirty-seven years on top of a pillar near Aleppo, Syria. His biographer relates that one of the saint's doubters tested his chastity by hiring a prostitute. For an entire night, Simeon fought the temptation. Each time the woman came close, he'd hold a finger in a candle's flame. The sharp pain kept him from giving in to his lust. He did manage to resist, but by the next morning, he had no fingers left.[14]

This apocryphal story seeks to illuminate sexual desire. Customarily, the male sex drive is presented as being so powerful that it is nearly impossible to control, easily triggered by straightforward visuals. Female desire, by contrast, is said to be fluid, contextual, and cycle bound.[15] Thanks to the unrelenting male drive, some men sire huge numbers of offspring. Famous human examples range from the Mongol conqueror Genghis Khan to the sultan of Morocco, Moulay Ismail "the Bloodthirsty." There is even a self-help book entitled the *Genghis Khan Method for Male Potency.*

The same may hold for other animals. Diego, a giant tortoise, single-handedly saved his species from extinction. As one of the few surviving representatives of his kind, he was moved from an American zoo to a breeding program on the Galápagos Islands, in Ecuador. Diego's unrelenting mating efforts helped raise the number of these tortoises from just fifteen to two thousand. One hundred years old, Diego keeps going.

At Burgers' Zoo, before releasing the chimps in the morning, I often visited their night quarters. When one of the females in the colony was sexually swollen, I could see it in the gleam of the males' eyes. Kept separate from the females at night, they were nevertheless keenly aware of the exciting prospect and were impatient to go outdoors to hang around her the whole day. They registered very little of what was going on around them. I could wave a banana in front of them, but they'd barely notice. Male chimps can go for days without eating when they are in this sex-obsessed state. For them, sex has priority over food. A female, on the other hand, may keep nibbling on whatever she was eating in the midst of intercourse.

The sexual stamina of male primates can be astonishing. The world champion must be a male stumptail macaque, who completed fifty-nine matings in six hours, each with an ejaculation. Although not at this extreme, male chimpanzees achieve high rates, too. The British primatologist Caroline Tutin observed more than one thousand copulations in the wild in Tanzania. Some males ejaculated on average once per hour, and younger males did so more often than older ones. In many primate species, males masturbate more than females and seem ready for sex anytime.[16]

In our species, an oft-quoted line has it that men think about sex every seven seconds. Although men, especially at a younger age, surely have sex on their mind much of the time, this number seems ridiculous. It would amount to eight thousand thoughts per day! The source is probably an old study by the Kinsey Institute, which found that most men think about sex every day, whereas most women don't.

Not everyone is convinced that men and women differ so greatly, however. Since recent scholarship suggests that the sex drive of women may match that of men, what actual evidence is there for a difference?[17] In 2001 three American psychologists published a comprehensive academic review on this issue. The lead author, Roy Baumeister, relates that he and his co-authors disagreed before they gathered the evidence. Kathleen Catanese held to the feminist "party-line," as Baumeister termed it, by predicting no difference. Kathleen Vohs was undecided, and Baumeister himself suspected the male drive was stronger. The three of them set out to scour hundreds of scientific reports for data on the sexual thoughts and behavior of men and women. Their assumption was that a stronger drive should express itself in more erotic fantasies, more risk-taking for sex, seeking more partners, suffering more from the absence of sex, and engaging more in masturbation. Sexologists often consider the latter the purest libido measure, as it doesn't depend on partner availability or the fear of pregnancy or disease.[18]

On nearly a dozen measures, without exception, men showed the stronger drive. Even though cultural disapproval of masturbation specifically targets boys and men (threatening them with blindness and insanity, no less!), men masturbate more than women. Also, men report that they find it harder to go without sex for long stretches. This even holds for those who—like Saint Simeon—have taken a sacred vow of chastity. Catholic priests break their celibacy more often than do nuns. As Baumeister cheerfully summed it up in his blog, "It's official: Men are hornier than women."[19]

Nevertheless, much of what is said about the female sex drive may need revision. Society sets such different moral standards for both genders that human studies, including those reviewed by Baumeister, can't be accepted at face value. Our double standard attaches negative labels to women who have casual sex, such as *slut, tramp, whore,* and *floozy.* These labels express sharp disapproval. In comparison, the labels for men who have sex with multiple women, such as *womanizer* and *skirt-chaser,* often come with a wink.

For researchers who seek to get around society's prejudices, the biggest obstacle is the social sciences' reliance on questionnaires. Especially regarding a sensitive topic such as sex, self-reports are hard to take seriously. No one wants to come across as a pervert or a jerk, so certain kinds of behavior are automatically underreported. Other kinds are overreported. Sometimes the data are plainly implausible. Thus, it is well known that men have more sex partners than women. And not just a few more, but more by a wide margin. One American study, for example, put the average number of lifetime partners of men at 12.3 and that of women at 3.3. Other countries report similar numbers. How is this even possible? Within a closed population that has a 1:1 sex ratio, there is no way. Where do men find all those partners? Many scientists have broken their heads trying to solve this riddle, but the most innovative approach tackled the likely source of the problem: lack of candor.[20]

At a midwestern university, Michele Alexander and Terri Fisher connected students to the tubes of a bogus lie-detector apparatus and asked them about their sex lives. Under the illusion that the truth would come out, the students gave very different answers than they had given before. All of a sudden, the women remembered more masturbation and more sex partners. On the first measure, they still scored below the men, but not on the second. Now we understand why the number of reported sex partners differs between the sexes. Men don't mind talking about them, whereas women keep the information to themselves.[21]

A PARALLEL DEBATE HAS been raging in animal research, where similar biases have been at play—although not, thank goodness, because we rely on questionnaires.

It harks back to the most fundamental sex difference of all, the one that biologists use to *define* the sexes. Our criterion is neither the look of an organism nor the shape of its genitals but the size of its reproductive cells, known as gametes. Gametes come in two varieties. Large ones are known as eggs, and individuals who produce them are known as females. Small, often motile gametes are called sperm, and the individuals who make them are known as males. In humans, eggs are one hundred thousand times larger than sperm, which is why scientists call sperm cheap and eggs expensive.

On top of this, female mammals have a long gestation period and nurse their young, whereas males contribute less and sometimes nothing at all. As a result of this difference in parental investment, the rules for maximizing offspring are different for the sexes. For a female, the maximum number is limited by what her body can accomplish. For the male, in contrast, his body needs only to produce sperm. For him, the limiting factor is how many females he can fertilize. Thus, males can be far more prolific than females. To put it in human terms, a man who has sex with a hundred women can, in principle, have a hundred children. A woman who has sex with a hundred men will still get only one baby at a time, rarely more. During her lifetime, she will have a limited number of children.

Evolution is driven by numbers of descendants: the more, the better. With this in mind, science decided that the above sex difference should lead to promiscuous males and discriminating females. Males will be eager and profligate, trying to fertilize as many partners as they can. Females will be choosy and reticent to make sure that they conceive with the best-quality males. This evolutionary rulebook is known as Bateman's Principle. It was formulated in 1948 by the English geneticist and botanist Angus Bateman, who backed it with experiments on fruit flies. Female flies produced the same number of offspring regardless of how many males they met, whereas males increased their number by meeting more females. Bateman's Principle is still the gospel of behavioral sex differences in nature, taught as undisputable to millions of students in biology and evolutionary psychology.[22]

So well-established and axiomatic are these ideas that we meet them everywhere in the literature on evolved human behavior. Here in the words of America's leading sociobiologist, E. O. Wilson: "It pays for males to be aggressive, hasty, fickle and undiscriminating. In theory it is more profitable for females to be coy, to hold back until they can identify males with the best genes. . . . Human beings obey this biological principle faithfully."[23]

This distinction between how the sexes play the mating game has lost its luster, however, especially with regard to females. The male side of Bateman's Principle is not at issue. There's good evidence that combativeness, like that of the zebra stallions, helps males win females. Males try to intimidate each other, jockey for status, push each other aside, or stake out a territory. Sometimes they kill each other, but most of the time, it's just a matter of winning or losing. Naturally, there are caveats: not every individual male is like this, and some males pursue alternative strategies, but overall this is how males spread their genes. The go-getter mentality of victors is inherited by their sons, who will propagate the same behavior. Men aren't exempt from this pattern, which has repeated itself for generation after generation for as long as sexual reproduction has been around.

The female pillar of Bateman's Principle, however, has begun to wobble and is ready to crumble. The whole idea that females are selective, chaste, faithful, and coy fits almost too nicely with our cultural prejudices, such as the widely held view that women are better suited for monogamy than men. This cliché struck many as so obvious that there was no need for critical examination. As a result, we don't have nearly as much information on it as we have on the male pattern.

Things changed only when ornithologists, instead of counting how many eggs their birds laid, began to determine who had fertilized them. Having discovered that female birds are quite the sexual entrepreneurs, they concluded that monogamy is mostly a surface reality. A distinction between genetic and social monogamy came into fashion, with most birds only having the latter. If the bird work put a dent in Bateman's Principle, it also didn't help that Gowaty was unable to replicate his fly experiments. Using improved methods, she failed to get the same outcome and has argued that his work was deeply flawed. As a result, his celebrated principle doesn't seem so persuasive anymore.[24]

This is where the primates come in. Because in their case, too, females refuse to fit the pattern.

IMAGINE ME SITTING in a corner reading my emails, minding my own business. All of a sudden, a woman rushes up to me. After some flirtatious eye contact with raised eyebrows, she pokes me in the chest or slaps my face. She's neither gentle nor subtle. Having gotten my attention, she then hurries off. After a short distance she stops and looks over her shoulder with a wide-eyed stare to check if I'm running after her.

This is how a female capuchin monkey solicits sex from the alpha male. I've worked with a colony of around thirty of these little monkeys for decades. We were always delighted by their courtship. It's like a playful dance in which the typical roles are reversed. They go through these poke-and-run charades the whole day until the male drops dead, or nearly so. During every copulation, both sexes whistle, chirp, and squeal excitedly. Males, however, are sometimes reluctant to the point of seeming indifferent. Or perhaps we should turn this around and say that the female's fire burns hotter than the male's, who has trouble keeping up.[25]

Whereas for male chimps, sex takes precedence over food, for male capuchins, the order is reversed. I have witnessed similar scenes on field visits to Brazil and Costa Rica. The American primatologist Susan Perry describes the following adamant wild female:

> *Capuchin males seem more interested in food than in sex in many cases, and we have even seen alpha males slap females who are pestering them for sex. One frustrated adolescent female, desperate for attention from the alpha male, bit him on the tail and pushed him out of the tree when he persisted in eating instead of responding positively to her advances.*[26]

Even though female capuchins definitely aren't coy or chaste, they do seem selective. Their courtship is directed mostly at the established alpha male, perhaps because he is the best male around. Females show a strong preference for a good male to lead their group. They support alpha males who protect them and keep order without being overly aggressive. Despite younger males being around, wild alpha males enjoy incredibly stable tenures, sometimes up to seventeen years. In our colony, I have seen incidents that illustrate the female role. One day our longtime alpha male lost his position to a junior upstart. We didn't witness the fight, but the young male must have either attacked the alpha or defended himself very forcefully. The alpha had deep lacerations (indicative of male canine teeth) and acted subdued. For three days, the females groomed him and licked his wounds. On the fourth day, he staged a comeback with their help, receiving massive support. His challenger stood no chance.

While capuchins seem to fit the "best male" hypothesis for mating, other primates fit the "many males" hypothesis. This idea comes from another prominent Darwinian feminist, the American anthropologist Sarah Blaffer Hrdy, who is the architect of an alternative view of female mating inspired by her fieldwork on Hanuman langurs. These elegant monkeys, named after the Hindu monkey god Hanuman, are found all over India. They are sometimes trained as primate police against the urban invasion of rhesus monkeys. With their black faces and menacing teeth grinding, male Hanuman langurs look intimidating. Twice the size of macaques, langur squads are effective at keeping these monkeys away from office buildings, garden plots, and the hallowed halls of parliament.

Langurs live in large troops led by a single adult male. The females mate with this male, but they also secretly undertake what Hrdy describes as "adulterous solicitations." They sexually invite males on the outskirts of the territory by presenting their rump while frenetically shuddering their heads. It's an unmistakable signal that invites copulation. These contacts are not risk-free, though. If the resident male catches a female in such an act, he will chase and slap her while herding her back into the fold. For young females, these liaisons might be a way to avoid mating with an alpha male who could be their father. But there is more to it, because not all cases can be explained this way.[27]

As Hrdy described it in *The Langurs of Abu,* she began to think in a new direction. For females, mating may be about more than getting pregnant. It may also be about ensuring the safety of their young. Males can be both helpful and harmful in this regard. We obviously expect them to be nice to youngsters that they have fathered. But remember, male primates have no idea of paternity. Instead, nature may have implanted a simple rule of thumb in their heads that does the trick. The rule might go as follows: *Tolerate and support the offspring of females with whom you have had sex in the recent past.* This rule doesn't require lots of brainpower or awareness of reproduction. All that is needed is a good memory. Males who follow it would automatically end up favoring potential sons and daughters.

Since infant care is almost entirely a female affair in langurs, male support mostly takes the form of protection. For example, during Hrdy's field season, a langur infant had been shocked and killed by electric wires in the bazaar of the nearby town. Its mother had not witnessed the incident. For over half an hour, the alpha male of the troop stood guard over the corpse, not letting any people come near, until the mother came to retrieve the body. Days later, when the mother temporarily left the body behind, Hrdy tried to inspect it, but the alpha male charged her.

She had to throw her notebooks and pen at him while scurrying away. Langurs have many predators (leopards, hawks, dogs, even tigers), and the large males are more effective than females in keeping them at bay.

But even more important is the "tolerate" part of the above rule of thumb. Langur males sometimes harm the young and not just a little. They turn murderous whenever they take over a troop. Any outside male that ousts the resident boss poses a grave threat to the young. This phenomenon, known as infanticide, is well studied. I happened to be present at the 1979 International Congress of the Primatological Society in Bangalore, India, where one of the first observers of infanticide reviewed his discoveries of earlier years. The pioneering Japanese primatologist Yukimara Sugiyama explained that he had seen a wild male langur snatching infants from their mothers' bellies, impaling them with his canine teeth.[28]

Sugiyama's lecture was the only one I have ever attended that got no applause whatsoever. It was met with deafening silence. The session chair took a patronizing tone and said that we had just heard an intriguing case of "behavioral pathology." Whereas Sugiyama had wondered, "Why does the new leader male bite all the infants?" his audience was not ready for this question. Infanticide is so awful that people don't want to hear about it. No one believed that Sugiyama's observations could have been more than a fluke. I still feel embarrassed by how his monumental discovery was received, especially given what we know now.

Hrdy reported similar incidents, saying that male langurs are laser-focused while stalking a female with infant. They circle her for hours like a shark while uttering a distinctive hacking vocalization before they attack. It looks entirely deliberate. Despite these observations, reports of langur infanticide were for decades met with controversy and shouting matches at conferences. Remember, this was long before we knew of all the other examples in the animal kingdom, such as the well-known case of lions. Langurs were the first species for which infanticide was described. The behavior didn't make any sense to most scientists, so it couldn't be true. Gradually, however, the reports became too numerous to ignore. They began to include other species, ranging from bears and prairie dogs to dolphins and owls. Male infanticide is now widely recognized.

The evolutionary explanation for this shocking behavior is that a new male can advance his own reproduction by eliminating the young of his predecessor. Once nursing cubs, pups, or infants are out of the way, females soon become fertile again. As a result, the incoming male can start fathering offspring earlier than otherwise possible, which gives him an advantage over males who fail to show such behavior. Sugiyama had an inkling of this explanation, and Hrdy elaborated on it further. In doing so, however, she didn't forget female counterstrategies. Whatever the males gain, infanticide is invariably devastating and harmful for the mothers. You'd expect them to try to thwart it, but how?

The key may be the above rule of thumb followed by males: *Tolerate and support the offspring of females with whom you have had sex in the recent past.* If this rule keeps males from harming offspring they may have sired, it also offers an opening for new mothers. All they need to do is mate with many males. If this fools the males into being nice to her offspring, she insulates herself from harm. Langur females, for example, could do so through contact with males who pose a future risk, such as those hanging around the troop's edges, waiting to take over. In other species, females may achieve the same by mating with multiple males. This is the crux of Hrdy's "many males" hypothesis.

Chimpanzee females seem to follow the many-males strategy. When a female shows up in the forest with a genital swelling, she attracts a large following. Several adult males trail her and mate with her in turn, one after the other, for the whole day. For wild chimpanzees, these gatherings can get quite large if there are several females swollen at once. These festive "sexual jamborees," as they have been described, take place without too much competition. At Burgers' Zoo, I spoke of "sexual bargaining," because the atmosphere was one of intense negotiation. Males would gather in clusters in the vicinity of the female, all grooming one another. They would allow one among them to mate undisturbed in exchange for a lengthy grooming session, especially with the alpha male. Every copulation had its price.[29]

When female chimpanzees enter the final swelling phase, competition among the males ramps up. She is most fertile at this stage. A high-ranking male will try to lure or force her away from the scene to have her all to himself. The main point, however, is that females copulate far more frequently and with more males than you'd expect if conception were her sole objective. It has been estimated that a wild female chimpanzee engages in six thousand matings with more than a dozen males during her lifetime. Yet she'll produce all together only five or six surviving offspring. If this sounds like an excessive amount of sex, it is—at least from a fertilization standpoint. It is not excessive, however, if we assume that females are trying to get sexually close to lots of males so that these males will leave her alone when her baby arrives eight months later.[30]

Chimpanzee males are infanticidal. By the latest count, over thirty incidents have been observed in four different wild populations, sometimes even including cannibalism of killed infants.[31] Naturally, human observers find this behavior repulsive. One Japanese fieldworker couldn't keep herself from stepping in:

> *Mariko Hiraiwa-Hasegawa observed several males surround a female who crawled on the ground and concealed her infant, while she pant-grunted (a submissive vocalization) fervently. Nevertheless, the villainous males attacked her one by one and seized the infant. On seeing this, Hasegawa momentarily forgot her position as a researcher and, brandishing a piece of wood, she intervened and confronted the males to rescue the mother and infant.[32]*

This is where bonobo females have a leg up. They have enormous amounts of sex, enough to include every male in the neighborhood and adjacent territories. Female bonobos pursue sex so actively and ardently that they almost turn coercive. Of all the primates I know, they are sexually the most forward. No killing of infants by males has thus far been observed. I regard bonobo society, with its widespread sex and sisterly solidarity, as the primate world's most effective female counterstrategy to male infanticide.[33]

PARADOXICALLY, WHAT WE ADMIRE most in nature is often tied to suffering. We appreciate the sight of mighty predators while forgetting how they make a living. We listen to the lovely call of the cuckoo at dusk while being blind to its cruel brood parasitism. Nature's dark underbelly stays out of sight most of the time. What better example than the vibrant sexuality of females, which may have evolved as a shield against male brutality. Not as a conscious tactic, of course, but as the reason why females endeavor to seek sex with more than one male. Their immediate motives are attraction, excitement, adventure, and pleasure. Behind the veil of evolution, however, we find the long-term improvement of offspring survival.

Our species isn't so different. Women, too, engage in sex far more often and with more partners than is strictly needed for pregnancy. Their immediate motivations may be richer and more varied than in other primates, but this still doesn't answer the question why women act this way. Evolution could have designed women to be sexually reserved, indifferent, and aloof, but clearly it didn't. Women routinely violate Bateman's Principle along with their marriage vows.

Hrdy applies the same evolutionary logic to human behavior. The special addition in our case is that we have a nuclear family structure. Men are involved with and provide for the young far more than do male apes. We have increased mutual dependency between the sexes. If a woman in a hunter-gatherer society loses her husband, she is in serious trouble. Her children will be in danger of being underfed. Binding men to her through sex is not only a way to avoid harm, therefore, but also a survival tactic linked to ensuring food and shelter.

On the danger side, it is good to realize that we are by no means free of infanticide. The Bible describes Pharaoh ordering children to be killed at birth, and most famously, King Herod "sent forth, and slew all the children that were in Bethlehem, and in all its borders from two years old and under" (Matthew 2:16). The anthropological record shows that after raids or warfare, the children of captured women are commonly killed. Hrdy has documented many examples of such behavior in grisly detail, which I won't repeat here. At minimum, we have every reason to include our species in discussions of infanticide by males.

Nor are we free from it in modern society. It is well established, for example, that children are far more at risk of abuse and homicide by stepfathers than by biological fathers. This suggests that men, too, take their sexual history with the mother into account. Besides, men know the connection with paternity.[34]

On the care side, we have the example of human societies in which children have multiple fathers. For example, children of the Barí in the Maracaibo Basin in South America often have one primary and several secondary fathers. The semen of all the men the mother has sex with is thought to contribute to the fetus's growth, a phenomenon known as "partible paternity." A pregnant woman will routinely take one or more lovers. On the day she delivers, she will utter the names of all these men. A woman who attended the birth will rush to the longhouse to congratulate each one of them, telling them, "You have a child." Secondary fathers have an obligation to help the mother and her infant. Survival into adulthood is higher for children with extra fathers than it is for those without.[35]

In most cultures, however, women don't benefit the same way from sleeping around. In modern society, we do everything to clarify paternity and prevent possible confusion. Our evolutionary history may not always have been so patriarchal, though. Although matrilineal and polyandrous societies are rare today, they may have been more prevalent in the past. The sexual adventurism of women in our species, even if it stays hidden most of the time, may have evolved for the same reasons as in our fellow apes. It may be an unconscious self-protective strategy to enlist male help and ward off hostility.

Female sexual preferences often deviate from the way males have arranged the mating system. When it comes to who mates with whom, a definite conflict of interest exists between the sexes. As Hrdy put it, "The breeding system best suited to the goose will often look different from the one preferred by the gander."[36]

All in all, it's time to abandon the myth that males have a stronger sex drive and are more promiscuous than females. We let this myth seep into biology during Victorian times, when it was enthusiastically embraced as normal and natural. We bent reality to meet our moral standards. This myth is still standard fare in biology textbooks, but support for it has never been overwhelming. Contradictory evidence for female sexuality has been accumulating with regard to both our own species and others. Female sexuality seems as proactive and enterprising as that of males, even if for different evolutionary reasons.

THE ISSUE OF female initiative came up again in relation to Diego, the tortoise with the legendary libido. If it hadn't been for him, his species would by now be extinct, we were told. But later we learned that Diego fathered only 40 percent of the offspring in the breeding program. A second male, with the uninspiring name of E5, apparently did the hard work. According to the American biologist James Gibbs, who ran the paternity tests, Diego got all the attention because he has "a big personality—quite aggressive, active and vocal in his mating habits." Noting that E5 was quieter yet more successful, he added, "Maybe he prefers to mate more at night."

My guess is that the female tortoises also had something to do with it.[37]

8

VIOLENCE

Rape, Murder, and the Dogs of War

After the attack on chimpanzee Luit, with which I opened this book, the zoo veterinarian tranquilized him and took him into surgery. He sewed hundreds of stitches while I handed him the instruments. We weren't prepared, though, for the gruesome discovery that we made during this desperate operation.

Luit's testicles were gone! They had disappeared from the scrotal sac even though the holes in the skin seemed small. Keepers later located them in the straw on the cage floor where the fight had taken place.

"Squeezed out," the vet concluded impassively.

Luit had lost so much blood that he never came out of anesthesia. He paid dearly for having stood up to two other males, frustrating them by his steep ascent in the hierarchy. The other two had been scheming against him, grooming each other every day. They took back the power they had lost. The shocking way they did so opened my eyes to how deadly serious chimpanzees take their politics.

If there is one aspect of social life that is gender-biased, it is physical violence. Males are its overwhelming source. This is universally true for humans (look up the homicide statistics of any nation), and it applies equally to most other primates. Not that female primates never turn violent, but they are more often on the receiving end. Males are victims, too, but generally at the hands of their own sex. Male brutality relates either to dominance and territoriality, when it is aimed at other males, or to sexual relations, when it targets females.

Evolutionarily speaking, competition over status and resources is the original reason for male aggressiveness. Human data reflect this. An extensive survey by the U.S. Justice Department estimates that annually 3.2 million men and 1.9 million women are physically assaulted.[1] Needless to say, most of these assaults are carried out by men. So I will start my exploration of violence by considering lethal combat among chimpanzees and the horrors of human warfare, both of which are predominantly male to male.

As the above numbers demonstrate, however, male aggressiveness is by no means limited to male adversaries. Men often exploit their size and strength advantage over women to molest them. In our societies, awareness and concern about femicide and spousal abuse is rising. Most violence against women stems from intimate partners. According to the Justice Department survey, 22.1 percent of women compared with 7.4 percent of men suffer such violence during their lifetime.

These figures no doubt underestimate domestic violence, which in addition includes rape. About one in six American women has been the victim of an attempted or completed rape. In reviewing what we know about violence between the sexes, as I'll do later on, the human species stands out. The incidence of such violence is higher for us than for most other primates. One possible cause is that human couples tend to live together in relative isolation. Our family arrangements facilitate male control and abuse, differing strikingly from the free-moving lifestyle of other primates.

Male-to-male competition within a group, such as what happened to Luit, is well documented in chimpanzees. Similar assassinations are known from chimps in the field. More often than not, they include emasculation by ripping off a rival's scrotum. Luit's injuries were not as atypical as they appeared to us at first. Male attackers often deliver the ultimate blow to another male's reproductive potential.[2]

I once spent time at the Mahale Mountains, on the shore of Lake Tanganyika in Tanzania, where my late colleague and friend Toshisada Nishida had followed chimpanzees since the 1960s. Nishida began his research at a time when science had no clue about the violent nature of chimpanzees. Our close relatives were still regarded as peaceful frugivores, a bit like Rousseau's noble savages. Since chimps were often encountered alone in the forest, or in ever-changing small groups known as "parties," they were thought to lead self-sufficient lives free from social ties. Nishida, however, noticed that they form distinct communities. This wasn't an easy discovery. To recognize that all of them belong to the same community required knowing every individual and keeping track of his or her travels.

Nishida's groundbreaking insight upset not only Western notions but also the expectations of his Japanese teachers. They were convinced that apes, like humans, would form nuclear families. When his professor arrived by boat for a visit, Nishida couldn't wait for him to set foot on land. Across the water, he shouted at him that there was no hint of nuclear families in our closest relative.

Nishida had been a great admirer of the legendary chimpanzee alpha male Ntologi, whom he called an "unparalleled leader." Ntologi managed to stay in power for an extraordinary fifteen years. He was a master at divide-and-rule and bribery, which he accomplished by freely sharing monkey meat with males loyal to him while withholding it from his rivals. But despite his political acumen, this legendary male was eventually brought down and expelled. He was forced to spend time on his own in the periphery of the community territory, barely able to climb, licking his wounds.

Ntologi did not show his face again until he could walk reasonably well. He'd appear in the middle of a social gathering and perform a spectacular display of strength and vigor. It was almost like the old days, when he was still in charge. But he'd revert to his limping, wound-licking existence as soon as he was out of sight. It was as if he used brief public interludes of stoicism to dispel any notions his rivals might entertain about his condition—a bit like how the Kremlin in the Soviet Union would parade its moribund leaders on television.

After making several comeback attempts, Ntologi one day returned as a broken male. He was forced to accept the lowest of low positions in the hierarchy. Two months later a gang of males attacked him. The scientists found him in a coma, covered with grave injuries. Nishida and his wife tried to revive him overnight at camp, to no avail. Ntologi died in the early morning.[3]

Even more common than these within-group battles is the unimaginable brutality of chimpanzees against outsiders. Not only do they form communities, but intense hostility prevails between them. Male chimpanzees regularly patrol the boundaries of their territory. They stalk victims across the border while staying completely silent to surprise them in a fruit tree. Several males (sometimes up to a dozen) overwhelm a single one in a highly coordinated attack. They bite and beat their enemy to a pulp, twisting his limbs until he is incapacitated, then leave him dying or dead. Sometimes they return days later to the exact same spot in the forest, searching for his corpse as if to make sure they killed him.

The first detailed report of this kind of "warfare" was presented in 1979 by Goodall, who described the systematic annihilation of one entire chimp community by another. This drama unfolded at Gombe National Park, not far from Mahale. It permanently shattered the peaceful image of the species. It was one of those discoveries that no one had seen coming, thus belying Donna Haraway's insulting suggestion that primatologists go to the field only to confirm their prejudices. If this were true, we'd still view apes as noble savages. Even Goodall herself was unprepared for what she had learned, saying, "It was a very dark time for me. I thought they were like us, but nicer."[4]

It took more than three decades for hard data to come in. A 2014 review in *Nature* listed 152 lethal attacks that had been observed or inferred in eighteen different chimpanzee communities throughout Africa. Almost all the assailants were male (92 percent), as were most of the victims (73 percent), and the majority of incidents were territorial (66 percent). I should add, though, that a relatively small number of communities were responsible for the majority of cases. Not every chimpanzee population has such a high incidence of violence.[5]

Luit was killed one year after Goodall's report. It shocked us because, at the time, we thought that only strangers would damage each other like this. Now we know better. The incident deeply affected me and my career. I decided then and there to devote my work to discovering what permits primates to live together. This was my emotional way of coping with an incident that had given me nightmares. I became a specialist in the way primates make peace after fights, cooperate, empathize, and even display a sense of fairness. Instead of despairing at the levels of aggression apes are capable of, my main interest became the ways they overcome these tendencies. Most of the time primates get along, including chimpanzees. Even though I never close my eyes to violence, and I realize how common it is under certain circumstances, it holds zero appeal for me. I am baffled by its glorification in movies and video games featuring gratuitous bloodshed.

In humans, too, violence is heavily gendered. The figures for humans and chimpanzees are strikingly similar. Of the world's nearly half a million homicides in 2012, men were the victim in 79 percent of the cases. Men were also the most common perpetrators, with a murder rate nearly four times as high as that of women.[6] And these numbers don't even include warfare, which adds another enormous male bias. Born in Europe right after World War II and familiar with its destruction, I have always looked at war as the great neutralizer of male privilege. Don't get me wrong; I am not one of those men who worries that our status in society is diminishing. I consider myself doubly privileged, by background as well as by gender. But I've been lucky. I came into this world during peacetime, and miraculously it stayed that way. Not entirely, of course, but the number of battle deaths in the world has steadily declined as armed conflict has moved from the international arena to within-state conflict.

Male privilege has always been most pronounced in the upper echelons of society. In the lower classes, men and women are equally exploited, mistreated, and impoverished. Had I been born fifty years earlier in a working-class family, my story would have been different. The outlook for poor boys was dismal. Being born male meant having a high chance of being drafted into the army and ending up riddled with bullets on some muddy battlefield. In the Middle Ages, death would have come by arrow, sword, or lance. Throughout history, the destiny of millions of young men has been an undignified and premature exit from life.

Boys were groomed for this fate. This is why, in retrospect, I have mixed feelings about my years as a Boy Scout. It all seemed innocent enough, but we did an awful lot of saluting, lining up in drills, stamping feet, and earning badges. The military ethos was considered good for a boy's character, but at the same time the Scouts' motto "Be Prepared!" related very much to war. By promoting discipline, teamwork, and conformity, the Scouts essentially molded boys into cannon fodder. The Shakespearian dogs of war were always begging to be fed. As Pink Floyd sang, the dogs don't negotiate, they don't capitulate, because "they will take and you will give, and you must die so that they may live."[7]

In modern times, we tend to forget this sad and distressing history of maleness. Every boy could be called upon to make the ultimate sacrifice. Objection was not only "unmanly" but a criminal offense. And power was always in the hands of older men. U.S. president Franklin Roosevelt once put it succinctly: "War is young men dying and old men talking." No nation would ever march one or two hundred thousand women to probable slaughter by the enemy. But young men were deemed of little value. Graveyards with endless rows of white crosses testify to the carnage. From the cynical (and Darwinian) perspective of older men, women are assets to be kept near and safe, whereas young men can be sent off to perish in distant lands for questionable causes. They are expendable.

Since warfare is mainly a male affair, its targets are often male, too. During the 1994 Rwandan genocide, enemy troops were looking to harm men and boys to such a degree that women tried to conceal Tutsi men by lending them their clothes. As one Tutsi woman described the lethal roundups, "They took all the men and boys, everyone masculine from about the age of two. Any boy who could walk was taken." The 1995 Srebrenica massacre, too, explicitly focused on Bosnian teenage boys. Nearly eight thousand of them were killed through summary executions. Generally, wars cost far more male than female lives.[8]

The killing of women doesn't come nearly as easily to us as the killing of men. In one experiment, American and British subjects preferred to sacrifice or punish a man over a woman. Asked whom they would hypothetically push in front of an oncoming train to save others' lives, nine out of ten participants of both genders preferred to throw a man rather than a woman onto the tracks. They offered reasons varying from "women are fragile, and it would be morally wrong" to "I value women and children over men."[9]

Under certain circumstances and to a limited degree, this bias allows women to shield others. Thus, in the summer of 2020 hundreds of mothers formed a human barricade in Portland, Oregon. They mobilized to defend demonstrators from armed federal officers, who had turned up to suppress protests in the city. The Wall of Moms—dressed in yellow shirts and mostly white—linked arms in front of the protesters, chanting "Feds stay clear! Moms are here!"

A real-life example of the reticence to harm women was seen during World War II. Nazi troops, who had shown no qualms about executing boys and men, began to revolt when ordered to do the same to large numbers of Jewish women and children. Even Adolf Eichmann couldn't fathom such horror, predicting that it would drive men insane. A solution had to be found. Note that the worry here was the troops' mental health rather than the fate of their victims. Gas chambers were considered the ideal option because this way the perpetrators didn't need to see their victims die. This method helped remove a major psychological barrier. Many historians suspect that without this monstrous innovation, the Holocaust couldn't have been extended to women, children, and the elderly. It would never have reached the scale it did.[10]

The sex-selective mortality of war has had long-term implications for gender relations. For example, the Soviet Union suffered staggering losses during World War II. The death of around 26 million people, most of them young men, thoroughly disrupted the marriage market. It resulted in a surplus of 10 million women compared to eligible men. As a result, men were in control of the postwar mating game and took sexual license. Large numbers of children were born out of wedlock. Moreover, the country absolved fathers of any legal obligations. Unmarried mothers were not even allowed to mention the father's name on birth certificates.[11]

Now that the world wars are long behind us, the ratio between men and women in most Western nations has climbed back to parity. The suspension of military conscription is an immense relief, but one that has come with unintended consequences. Male privilege has become more glaring. It isn't offset anymore by that awful sword of Damocles hanging over every young man's head. Able to walk around free of worry about the next possible war, we have begun to see more clearly how easy some men have it. Combined with a reduction in family size thanks to the Pill, these changes have put fresh bargaining power in the hands of women. Today's renewed gender debate flows from these demographic shifts in society.

WHILE SHE WAS following chimpanzees at Gombe National Park, the primatologist Barbara Smuts hit back at a male named Goblin. He was trying to terrorize her and steal her prized rain poncho. Goblin had been picking on Smuts for days until she decided that enough was enough. In the scuffle over her poncho, she instinctively landed an angry punch on his nose:

> *After I punched him, Goblin crumpled into a whimpering child and went to Figan, the alpha male, for reassurance. Without glancing up, Figan reached out and patted Goblin several times on the top of the head. I later realized that Goblin had been treating me just as he was treating some of the adult female chimpanzees.*[12]

As an adolescent, Goblin was at the age when males are busy intimidating females in their community to assert status. He was doing the same with Smuts. Smuts had tried to overlook his provocations but then learned from watching female chimps that "by ignoring Goblin, I had failed to send a clear signal." The best response was to fight back. After she landed one good punch, he didn't bother her anymore.

Male chimps show a second form of harassment, which relates more directly to sex. Chiefly directed at fertile females, it can be brutal. In Kibale National Park, in Uganda, they have even added weaponry. Males beat females with large wooden clubs. The first observation was an attack on a swollen female named Outamba. As fieldworkers watched, a top male, Imoso, held a stick in his right hand and hit Outamba about five times, beating her hard. This action exhausted the male, who took a break for a minute, after which he resumed the beating. Now Imoso was handling two sticks, one in each hand, and at one point he hung from a branch above his victim, kicking her with his feet. Finally, Outamba's little daughter came to her mother's aid, pummeling Imoso's back with her fists until he gave up.

Imoso's technique inspired copycats. Following his example, other males started doing the same. They always chose wooden weapons, which the investigators saw as a sign of restraint. The males could also have picked up rocks, but that might have harmed or killed their mates, which was not their goal. Their goal was to instill obedience.[13]

Does such behavior aid a male's reproduction? It's a question that evolutionary biologists ask of every typical behavior. There is indeed evidence that male chimpanzees who harass females sire more offspring, but how this connection comes about remains a mystery. Since copulation seldom follows the rough treatment of a female by a male, the connection must be indirect. It could be that aggressive males instill fear so that females will acquiesce at the critical moment, or perhaps these males produce more viable sperm. We don't know.[14]

A much more straightforward form of coercion would be rape, which is called *forced copulation* in animals to avoid the human connotations of the former term. For years, the FBI defined *rape* as "carnal knowledge of a female forcibly and against her will." This rather wishy-washy definition, however, suggests that only women can be raped. Since 2013, the FBI definition of rape specifies penetration of either the vagina or the anus without consent.[15] Applying this definition to other primates, the behavior we'd expect is that of a male restraining a female and penetrating her while she struggles to free herself. Thus defined, however, forced copulation is almost unheard of. It has been seen a few times in chimpanzees when sons or brothers tried to mate with a female, something every female strenuously objects to. One wild chimpanzee refused her son's sexual advances but eventually submitted when he kept harassing her. She did so under screaming protest and jumped away before he could ejaculate.[16] Other than in this intrafamilial context, forced copulation is exceedingly rare among chimpanzees. Even though I must have watched over a thousand couplings in captive chimpanzees, I have never seen sex against a female's will.

That female chimps know how to ignore or resist male harassment is clear from Nishida's lifetime of field observations: "When females show reluctance to comply with males' solicitations, adult males sometimes display aggressive behavior by bluffing, hitting, or kicking them.

Estrus females, however, stubbornly resist such violence and intervention and almost never succumb to it. Of 12 cases in which the two oldest males resorted to violence in response to females' refusal, none resulted in mating."[17]

This leaves us with a picture of chimpanzee society as definitely rough and abusive to females, especially fertile ones, but with questions about sexual coercion. When a female develops a genital swelling, she generally mates with various males without problems. Only when her cycle reaches its peak, during ovulation, do high-ranking males impose restrictions. Putting a halt to all this free love, they seek to convince her to go "on safari" with them. They may threaten or punish her to achieve this goal. While I was at Mahale, alpha male Fanana disappeared for two entire months with an estrous female. Keeping her away from other males, he monopolized her. Paternity analyses indicate that these consortships often result in conception.[18]

The Mahale chimpanzees live dispersed over the forest. Traveling alone or in small groups, they can't see each other most of the time due to the dense foliage. They are very much in tune with the sounds coming from everywhere, though, and they seem to know exactly where every other chimp is by the frequent, loud hooting for which the species is known. Chimpanzees recognize each other's voices. They often interrupt their travel to cock their heads and pick up sounds that may originate more than a mile away.[19]

During his safari, however, we never heard Fanana. He and his mate must have been traveling and foraging in complete silence, otherwise they'd quickly have drawn attention. How could he have kept her close against her will in dense forest for months on end? Since females sometimes protest against a safari by uttering rousing shrieks, her being silent for such a long time suggests consent. Or perhaps she suppressed her calls so as not to attract danger. Consorting pairs often travel close to the territorial boundary with hostile neighbors.

Fanana returned from his extended safari with loud calls and a most impressive charging display. He left no doubt that he was in perfect shape, ready to reclaim the top spot. In the intermediate period, the beta male had taken on the alpha position, but without being at ease. He was incredibly nervous about Fanana's return, so much so that we had trouble keeping up with him running up and down the hills. The turmoil of that day left me totally exhausted.

ALL THE ABOVE observations stem from East Africa. In West Africa, chimpanzee females aren't treated nearly as roughly. This seems to be a cultural difference.

Most stories one hears about chimpanzees concern the Eastern subspecies and habitat, where chimpanzee field research began in the 1960s. Here the apes live dispersed over the forest, territorial violence is common and severe, and females wield little power. This emphasis is unfortunate, since chimpanzees don't always behave this way, as I know all too well from my studies of them in captivity. The species has a great potential to form more cohesive and cooperative societies. Fieldwork in West Africa confirms this. Although clashes between communities are not absent, they are less frequent and less brutal. Western chimpanzees fail to support the murderous image of the species. Within each community, we find more social closeness and less of a power differential between the sexes.

This contrast with Eastern chimpanzees was drawn by the Swiss primatologist Christophe Boesch. For decades, he worked with the chimpanzees of Taï Forest in Ivory Coast, in West Africa. Boesch provocatively titled a book about his studies *The Real Chimpanzee,* to the great annoyance of chimpanzee experts working elsewhere in Africa. It was a bit like an anthropologist claiming to be the only one studying "real humans." But even if we don't accept that chimpanzees vary in authenticity, we should treat generalizations about their behavior with caution.[20]

The higher level of cooperation among Western chimpanzees may be due to the numerous leopards in the forest, which require collective defense. The chimps' togetherness has the side-effect of shifting the sexual power balance within communities. When females spend much time in each other's company, they form a bloc of common interests. This puts a halt to brutal male tactics. According to Boesch, females have a bigger say in community affairs and aren't subjected to forced sexual consortships and coerced mating. Moreover, if a female does mate with a less preferred male, she often darts away prematurely, thus preventing fertilization. Taï females have more control over their sexuality and by extension their reproduction.[21]

In captivity, male chimpanzees are even less successful at pressurizing females. You might think that the absence of escape opportunities would make it easy for them to intimidate females, but it's the opposite. The collective power of females exceeds that seen in the wild, because in captive colonies they are close together all the time. Social life is much more tightly regulated, and males can't get away with obnoxious behavior. I have seen males bluff with all their hair on end at females who are reluctant to mate, but there always comes a point when other females jump in to save the screaming victim. They go after the unrelenting male and teach him to behave.

The same pattern marks bonobos, who have turned sisterly solidarity into an art. They curb male violence both in captivity and in the wild, where their communities are remarkably tight. Female bonobos travel together most of the time and at nightfall call each other before building their nests high up in a tree cluster. They sleep within hearing range of each other. Bonobos have more togetherness than chimpanzees, resulting in an even greater transfer of power toward the females. Sexual coercion by males is out of the question.

What do these observations mean for the distant ancestor that we share with the apes, who lived between 6 and 8 million years ago? If you asked me if this ancestor was a rapist, my answer would be no. We have no reason to think so, given how extremely rare forced copulation is in our next of kin. Did this ancestor know at least sexual coercion in the form of harassment and intimidation? That would depend on how close-knit its societies were. We do have evidence for such behavior in chimpanzees who live dispersed over the forest. It's exceptional, how-ever, in more cohesive societies of the same species and totally absent in bonobos. Most sexual activity occurs in a relatively relaxed context. The only behavior we'd expect in this ancestor is the occasional battering of females by males, at least if we assume male dominance. Since we don't know to what extent the last common ancestor was chimp- or bonobo-like, this point remains up in the air.

Unfortunately, none of the above helps us explain the behavior of our species, in which rape is more common than most of us dare to admit. It is far more prevalent among us than in our primate kin. According to the same large U.S. survey I mentioned earlier, 17.6 percent of women are raped at some point in their lives.[22] This high number may be partly due to the tendency of human couples to spend time together in dwell-ings separate from the rest of the world.

In our ape relatives, males and females don't associate permanently, and they meet only occasionally. Most of the time, females travel freely on their own while foraging for themselves and their offspring. At night, they build a nest in the trees. The males are barely interested in how the females lead their lives. Outside the female's fertile periods, the sexes have no reason for frequent contact, least of all for abusive male control and jealousy. Given the absence of a family structure, there is far less close monitoring of the activities of one sex by the other. Moreover, their encounters take place out in the open, where others may interfere.

Our species evolved families marked by male engagement. The advantages of this arrangement in terms of food provisioning, protec-tion, and childcare are part of our species' success story. Yet these advan-tages came at an enormous price to women in terms of men's attempts at domination and control, including rape. Cohabitation sets up a situation that potentially endangers women. An amplification of this effect was seen in 2020, when the COVID-19 outbreak led to stay-at-home orders. With families cooped up even more than usual, domestic violence tri-pled in China's Hubei Province and other places. Preliminary reports suggest a worldwide increase in domestic abuse.[23]

To ALL THIS TALK about the infrequency of sexual coercion in other primates exists one giant exception: the orangutan. The red ape of Southeast Asia is much less close to us genetically, hence is less relevant to our ancestry than chimpanzees or bonobos, but everyone who knows orangutans has witnessed the behavior in question. The male grabs a female and doesn't let go. He has essentially four hands and is incredibly strong. He forcefully achieves intromission and thrusts while she struggles to get away. Halfway, she gives up and waits for it to end.

This behavior is aided by the fact that orangutan males are considerably larger than females. Orangutans' solitary lifestyle is also a factor. Instead of gathering in groups, these apes travel alone through the canopy, high above the ground. A female will usually be on her own, accompanied only by dependent offspring. The absence of a support network gives males the upper hand.

So strong is the tendency to copulate by force that orangutans are even known to target women. This happened to the cook of Biruté Galdikas, the Canadian primatologist who studied the species for decades in Borneo. One day an orangutan whom Galdikas had raised since infancy—he slept for a while in the same bed with her and her husband—grabbed her Dayak cook. He ripped off her skirt and forced himself onto her while she was "screaming hysterically." Galdikas tried to intervene but to no avail, because a male orangutan is many times stronger than any human. In the end, he released the woman without leaving any injuries.[24]

Not all orangutan males act this way, however, and females don't always resist sex. It depends on the male's status and size. Forced copulations are typical of smaller males, who lack secondary sexual features, such as the large fleshy cheek pads, or flanges, on the sides of their faces. These males often live in the territory of a fully grown male and mate with females even if the females don't want them. The females prefer the larger males, which are generally twice their size, have flanges, and regularly utter loud calls from the treetops. Their deep long calls can be heard far and wide. I have stood underneath them in the forest, and the force with which they announce their presence gave me goosebumps. Females are eager to have sex with these magnificent males. They actively seek them out and even orally bring about an erection or use their fingers to help insert their penis. Carel van Schaik, a Dutch expert on orangutans, describes the procedure:

> If a young female . . . wants to mate with the irresistibly attractive big, flanged, dominant male, she's got some convincing to do. Indeed she's got work to do: approach the bored-looking male and mount him while he is leaning back, achieve intromission, and thrust hard on him to get him to ejaculate.[25]

The reason for these opposites in mating isn't fully understood. Fieldworkers suspect that females resist some males and not others based on how much security a male can provide once their offspring is born. The largest males, who rule a vast stretch of forest, are no doubt the better protectors.[26]

What we do know, however, is that forced encounters rarely if ever cause injuries. Despite their enormous size advantage, male orangutans engage in this behavior without making visible wounds. Their bites must be mere threats. In general, the aggression of male primates is restrained when it comes to females. This is evident from their choice of weapons (wood versus rock), and the fact that chimpanzee males rarely kill females. Even when males encounter strange females in the forest, they often leave them alone. Their territoriality focuses on other males.

For gorillas, the story is similar. The male of this ape species is the most formidable fighting machine in the primate world, physically capable of holding off or killing a number of the much smaller females. Psychologically, however, he is incapable of fully exploiting this advantage. In confrontations with females, he mostly bluffs and beats his chest. It is quite spectacular to see an alliance of barking female gorillas chase— even beat—a colossal male, whose hands seem tied behind his back by the neurons in his brain.

Orangutans often use force during sexual intercourse. This is mostly done by males who are not yet fully grown. Females prefer sex with the larger males, who are twice their size.

These inhibitions make perfect sense. If reproduction is the primary objective of male aggressiveness, its lethal use against females would be about the most unproductive thing to do.

Rᴀᴘᴇ ʜᴀs ʙᴇᴇɴ used as a weapon of humiliation and terror against millions of women. Examples include the Japanese army during the 1937 Rape of Nanjing, the Soviet's Red Army in Germany at the end of World War II, and the Hutus during the 1994 genocide in Rwanda.

Rapes turn genocidal when they are followed by torture and killing, as they often are, or lead to botched abortions or deadly diseases, such as AIDS. Further back in history, troops are said to have raped and pillaged as a way of punishing cities for refusing to surrender. The Mongol leader Genghis Khan, for example, would give cities under siege an ultimatum: "Whoever submits shall be spared, but those who resist, they shall be destroyed with their wives, children and dependents."[27]

Nowadays, however, the most significant source of violence against women resides in their own homes: their intimate partners and family members, such as boyfriends, husbands, and brothers. Globally, an estimated 13.5 percent of all homicides are *femicides,* defined as sex-based hate crimes.[28] Rape is part of this worldwide pattern, but even though the reliability of the available numbers is much debated, the one measure we never get to see is how many perpetrators there are. Is it one in five men? Criminal records suggest that rapists are serial offenders, so it could also be one in ten. Or perhaps one in twenty. This is a critical question if we wish to determine which factors promote rape. Is rape typical of our species or an exceptional pattern that marks a small minority of men?[29]

The occurrence of rape is sometimes taken as an encapsulation of our species's gender relations. In her 1975 book *Against Our Will,* Susan Brownmiller wrote the memorable lines:

Man's discovery that his genitalia could serve as a weapon to generate fear must rank as one of the most important discoveries of prehistoric times, along with the use of fire and the first crude stone ax. From prehistoric times to the present, I believe, rape has played a critical function. It is nothing more or less than a conscious process of intimidation by which all *men keep* all *women in a state of fear.*[30]

In speaking of *all* men and women, Brownmiller made a sweeping generalization that offers little room for the role of culture and education. She also didn't distinguish between men who rape and those who don't. Her main point was that it doesn't matter how many men engage in this behavior since *all* women live in constant fear of it and are forced to take self-protective measures.

Knowing how many men rape is important for those who'd like to get rid of this behavior. Others, however, think this won't be possible. They consider rape natural to our species. Rape is neither an act of violence nor a cultural innovation, they argue, but an adaptive strategy. In their 2002 book *A Natural History of Rape*, the American scientists Randy Thornhill and Craig Palmer present rape as part and parcel of our evolutionary psychology. They see it as a preprogrammed solution for men to deal with sexually unwilling women. What they mean by "adaptive" is that rape helps men achieve fertilizations that they otherwise would miss out on.[31]

I fully understand Brownmiller, who, out of anger about the prevalence of rape and its traumatic impact, was ready to blame an entire gender. I have more trouble with the biologization of rape by Thornhill and Palmer, partly because of what we know about our fellow primates, and partly because of the scant evidence regarding our own species. Also, calling rape "natural" gives the impression that we just have to live with it. The authors' assurances that this is not what they meant never sounded convincing.

The idea that rape is an adaptation was born, believe it or not, from research on scorpion flies. Some fly species have a physical feature—a sort of clamp—that helps males force females into copulation. While it's a stretch to extrapolate from flies to us, the authors do their best. Men obviously lack an anatomical rape tool, but the authors speculate that perhaps their psychological makeup facilitates rape. The problem is that human psychology is not nearly as easy to pick apart as insect anatomy. The human species is far too loosely programmed for highly specific behavior, such as rape, to be heritable.

Advocates of the rape-as-adaptation view always trot out the handful of animals that engage in forced copulation, such as ducks and orangutans. Based on evolutionary logic, however, one wonders why these animals are the only ones. If rape is such a great impregnation technique, why would it be so rare? Forced sexual contact should be rampant in nature, but it isn't.

For natural selection to favor rape, two conditions would have to be met. First, men engaging in this behavior should have a special genetic makeup that turns them into sexual predators. Second, rapists would need to spread their genes. We have no evidence for either condition, however. Moreover, if reproduction were the goal, men shouldn't rape girls or women outside the reproductive age range. Nor should they rape lovers and wives, with whom they also have consensual sex, or boys and men. Yet they do. For example, according to the survey of the U.S. Justice Department, one in thirty-three men is raped during his lifetime.[32]

I was horrified by the prospect of such sloppy biology reaching a broad audience, but it did. A Natural History of Rape became a millstone around the neck of the young discipline of evolutionary psychology, which until then had mostly been known for harmless speculations about the attractiveness of hips, waists, and facial symmetry. The controversy culminated in a countervolume in which twenty-eight scholars rejected Thornhill and Palmer's thesis. Joan Roughgarden called it "the latest evolution-made-me-do-it excuse" for depraved behavior.[33]

In my critical review of the book for The New York Times, I brought up a different issue: what would a tribal community do with a rapist in its midst? I was thinking of our long prehistory in small-scale groups.[34] Led by Kim Hill, American anthropologists explored this question based on what they knew about the Aché Indians of Paraguay. They hadn't heard of any rapes among these hunter-gatherers, but built a mathematical model based on how they expected these people to respond if one of their men had raped a woman. It didn't look good. The rapist might lose all his friends or end up killed by relatives of the victim, while any potential offspring might be abandoned. A rape gene—if it existed at all—would probably quickly die out.[35]

Our nearest relatives don't show signs of a rape adaptation, and in the conditions under which our ancestors evolved, rape couldn't have been a wise move. In today's huge societies, anonymity removes some of the risks for perpetrators, but the fact that rape happens still doesn't make it natural.

Smuts was the first to speculate about how human society shapes male violence and sexual coercion and how it can be countered. In doing so, she was inspired by her observations of primates. As we have seen, the stronger the female network, the more male sexual harassment is being curbed. Female primates tend to defend each other against males, but for this to work, they need to live and travel together. The female orangutan, who has absolutely no backing, is in a perilous situation compared to the female bonobo, who has a first-rate support alliance.

To prevent harassment by men, women have three main options available, according to Smuts. One is to fight back, which is their first choice. But this is hard to do and dangerous as men are on average stronger. The second option is to find adequate male protection. This choice is open to many primates, too. Think of the male-female friendships of baboons and how orangutan females are drawn to the mightiest male in the forest. This tactic has drawbacks too, though. If a woman selects a mate based on his vigor and dominance, she runs the risk that he will turn these same qualities against her. A strong protector is potentially also a dangerous bully.

From a woman's perspective, the perfect man is strong enough to daunt other men yet gentle enough never to exploit his physical advantage against her. That heterosexual women are attracted to these qualities is clear from their pronounced preference for tall men. Women favor men who are taller than themselves to such a degree that short men complain that they stand no chance on dating services. Women's attraction to men they can look up to exceeds men's desire for women shorter than themselves. Asked about mate preferences, women insist more on the intergender height difference than men.[36]

Toughness is an additional factor. Women tend to appreciate well-chiseled abs. When Brad Pitt, standing on a rooftop in the sun, took off his shirt in *Once Upon a Time in Hollywood*, cinema audiences are said to have gasped. We are quick to gauge a man's abilities from his torso and arms. If we look at pictures of headless shirtless male bodies, we have no trouble ranking them on upper body strength. Tested on such pictures, women prefer muscular-looking torsos. A sample of 160 American women included not a single exception to this rule. Other studies have reported that women are turned off by excessive muscularity, but those studies featured cartoonish line drawings. Within a normal range, women overwhelmingly go for the health and strength of potential mates.[37]

Whenever unsavory men threaten women, the most effective help comes from other, better men. We find this solution so satisfactory that the male hero saving a damsel in distress remains one of our most popular fiction clichés. The hero needs muscle power, though. Meek men may be safe at home, but out-and-about women prefer the company of men who are able to stand their ground. Knowing that they are judged on this quality, men advertise it. This may be why they are drawn to competitive sports. All over the world, men watch and play sports more than do women. And women pay attention, as was demonstrated when they judged pictures of men accompanied by a statement about what sport they played. Women favored a man who was said to engage in an aggressive game, such as rugby, over the same man in the same picture who was said to play a gentler game such as badminton.[38]

The third way for women to reduce the risk of male sexual harassment is to rely on each other. Their support network may be kin-based (if women stay in their natal communities after marriage), but it could also, like the bonobo sisterhood, consist of unrelated women. The #MeToo movement comes to mind. So does the Green Sari movement.

In a small village in northern India, women privately endured frequent domestic violence from their drunken husbands. One day these women banded together and began to roam the streets of their village. They did so every evening, all dressed in green saris. They quickly became a force to be reckoned with, smashing up bootleg bottles and confronting men who were troubling their wives.[39]

Smuts developed a set of predictions about situations that would protect women, such as the proximity of kin, reduced dependence of women on men, and less emphasis on male bonding in society. Men who spend much time in men's clubs or brotherhoods shift their priorities away from women and become reluctant to defend female relatives against other men. These predictions have thus far not been tested with actual data on human cultures. Still, they offer an excellent outline of a cultural approach to the problem of sexual harassment and coercion.[40]

This framework strikes me as far superior to the assumption that rape is in our genes and that men will resort to it whenever the opportunity arises. The latter is a depressingly fatalistic position that denies that men can do better. This is why I'd like to add a fourth option to fight male harassment and rape: create a culture in which boys and men aren't drawn to such behavior and don't condone it in their friends. Instead of focusing on what women can do to prevent it, we also need to consider what we teach boys and what kind of models we provide them with.

The question I prefer to ask is why most men *don't* rape. Let's focus on the positive and see how this majority can be increased. Education will be critical, especially one that acknowledges sex differences. The American feminist Gloria Steinem's recommendation that we raise sons more like daughters will have to be taken with caution.[41] We can't act as if biology were irrelevant. Sons aren't daughters.

If the above descriptions of primate and human behavior teach us anything, it's that sons will grow up to be more prone to violence. They will also acquire considerably greater body strength than daughters. Every society needs to come to grips with this dual potential for trouble and find ways to civilize its young men. Since boys rarely become warriors anymore, society has an even greater need to find constructive outlets for their aggressive drive. This drive can lead either to great accomplishments or to bad behavior. To make sure they become sources of strength rather than abuse, boys need to acquire emotional skills and attitudes geared specifically to their gender. They need to learn that strength comes with responsibility. We want them to develop self-discipline, a sense of honor, and respect for women.

Not just as a side issue, but as the core of their masculinity.

9

ALPHA (FE)MALES

The Difference between Dominance and Power

Mama was the center and rock of the large chimpanzee colony at Burgers' Zoo. She acted as the mother of the group—hence her name. Mama reigned as alpha female for over four decades, dealing with multiple alpha males who came and went. Of all the top chimpanzee females I have known, Mama was the one with the best leadership skills. She cared not only about her privileged position in the hierarchy but also about the group as a whole.

Mama demanded so much respect that I felt small the first time I looked into her eyes at face level across the water moat. She had a habit of calmly nodding at you to let you know that she had seen you. I had never sensed such wisdom and poise in a species other than my own.

In later years, after I had left the country, Mama would enthusiastically greet me every time she spotted my face in a crowd of visitors. My visits were always unexpected and sometimes occurred after years. She'd jump up and run to the water moat's edge, hooting while stretching out a hand to me from a distance. Females typically use this "come here" gesture when they are about to move and want their offspring to jump onto their back. I'd make the same friendly gesture back at her and later help the caretaker feed the chimpanzees by throwing fruit onto the island. We'd make sure that Mama, who walked slowly and wasn't as quick at plucking flying oranges out of the air as the others, got enough.

Jealousy was also on display because Mama's adult daughter, Moniek, had a habit of sneaking up on me to lob rocks from a distance. Moniek's parabolic launches would have hit me in the head if I hadn't learned the hard way to keep an eye out for this sort of behavior. I have caught many a rock in the air! Moniek, born while I still worked at the zoo but not remembering me, hated her mother's attention for this stranger, whom she greeted like an old friend. So better throw something at him! Since aimed throwing has been claimed as a human specialization, I have invited proponents of this theory to discover what chimps are capable of. Moniek had perfect aim from over forty feet away. I have never had any volunteers to test their pet theory, though.

Within the group, Mama acted as the voice of collective opinion. A typical example concerned Nikkie, a brand-new alpha male. Although he had gained the colony's top spot, others resisted his heavy-handed behavior. Being the alpha male doesn't mean you can do anything you like, especially not for one as junior as Nikkie. One time all the disgruntled apes chased him, loudly screaming and barking. The young male, no longer so impressive, sat alone high up in a tree screaming in panic. All his lines of escape were cut off. Every time he tried to come down, the others chased him back up.

After about a quarter of an hour, Mama slowly climbed into the tree. She touched Nikkie and kissed him. Then she climbed down while he followed close at heel. Now that Mama was bringing him with her, nobody resisted anymore. Nikkie, obviously still nervous, made up with his adversaries. No other chimp in the group, male or female, could have brought about such a smooth closing.

There were many occasions when Mama brought quarreling parties together, or these parties sought her out. I have seen full-grown males, who couldn't resolve their fight, run up to Mama and sit, each in one of her long arms, screaming at each other, almost like ape children. Mama would keep them from starting the quarrel all over again. At other times, she'd encourage a male to approach his adversary and make up. Her behavior reflected a keen appreciation of the social dynamics around her. Her mediations reflected *community concern:* they went beyond immediate self-interest by promoting peace and harmony in the group.

Since she was always ready to do her part, others counted on her. Females who could not quash a commotion among juveniles would wake up Mama by softly poking her in the side. Juvenile fights risk stirring conflict among the mothers, who automatically side with their offspring, which would only make matters worse. The solution was to activate a neutral party of undisputed authority. Mama needed to utter only a few angry grunts from a distance for the young chimps to stop their shenanigans.

THE TERM *alpha male* goes back to wolf research in the 1940s done by the Swiss ethologist Rudolf Schenkel.[1] Soon after he coined the term, students of animal behavior began to use it to refer to the top-ranking male. The top-ranking female became the alpha female. Both males and females have their alpha, and there is never more than one of each in a group.

Mama did not dominate any adult males—female chimpanzees seldom do. This simple truth about most primate hierarchies has been a thorn in the side of some. What is there to learn from primatology, some feminists ask, if all it can deliver is a depressing message about women's place in society? On the other hand, conservatives celebrate the same information as a justification of "under my thumb!" male attitudes.

In 2013 one American pundit, Erick Erickson, declared on Fox Business television, "When you look at biology, the roles of a male and a female in society and in other animals, the male typically is the dominant role." He saw this as scientific proof that breadwinning women are a crime against nature. The only possible outcome of their rise, all the male members of the show's panel somberly agreed, is the breakdown of society.[2]

Primatological support for the message that women need to know their place goes back to Solly Zuckerman's ill-begotten baboon experiment at the London Zoo in the 1920s. His views helped justify, even glorify, male brutality. In the 1960s the journalist Robert Ardrey's influential book *African Genesis* amplified the message, including the following hostile (and phobic) statement about changing gender roles: "The emancipated woman of whatever nationality is the product of seventy million years of evolution within the primate channel. . . . She is the unhappiest female that the primate world has ever seen, and the most treasured objective in her heart of hearts is the psychological castration of husband and sons."[3]

Ardrey's worry about women's happiness was disingenuous. The underlying assumption was that male and female leadership are mutually exclusive and that the former is more natural than the latter. But what if the two coexist?

The reason we rarely hear about feminine power in other primates is that we can't look past male leadership. Males are flamboyant and suck up all attention with their cockiness, displays, and noisy fights. They are also less timid, which means that fieldworkers get to know them first. Prominent female primatologists haven't escaped their allure, developing a special rapport with charismatic male apes such as Jane Goodall with David Greybeard (a chimpanzee), Dian Fossey with Digit (a gorilla), and Biruté Galdikas with Sugito (an orangutan). These males were described with great love and admiration while female apes received less attention, at least initially. Given their low-key behavior, it took decades for them to enter the scientific literature.

Secondly, male dominance is associated with violence, which we are drawn to. We find it hard to look at anything else. This well-known bias in how we watch the daily news applies equally to animal behavior. Television offers far more nature programs about sharks than about mountain goats. Whenever I ask producers of nature series why we have a zillion documentaries on chimpanzees and so few on bonobos, the answer is always that the latter don't show enough action. While filming chimpanzees, camera crews are guaranteed to get a few spectacular fights. Blood and confrontation sell. The program can show a bleeding chimpanzee limping away from the scene while the narrator grimly reminds us of "the law of the jungle." Nature networks love to leave us with this dark message.

They have boxed themselves in, though, excluding more thought-provoking narratives. Bonobos do show plenty of action, although mostly of an erotic kind. The networks have trouble handling this. Furthermore, featuring bonobos could dispel the notion that male dominance is inevitable, which poses yet another problem. How could the law of the jungle ever have put females in charge? It's just too hard to explain, producers tell me.

Similar biases weigh down the scientific literature. Evolutionary scenarios of our species generally present us as born warriors who have raided, plundered, and killed since time immemorial. This grisly prehistory is thought to explain our most cherished traits. As summarized by the American political scientist Quincy Wright, "Out of the warlike peoples arose civilization, while the peaceful collectors and hunters were driven to the ends of the earth."[4]

Given that warfare requires a high degree of cooperation and mutual help, even human altruism is seen as an offshoot of militarism. Civilization and obedience to authority are thought to have evolved so that we could more effectively confront our enemies. Human anatomy is viewed in the same light. You might think that our hand evolved to grasp branches and pick fruits, but since a hand can also be balled into a fist, the latest proposal is that human hands are meant to be weapons.[5]

These views are reflected in the labels bestowed on the males of our species and its relatives by anthropologists, such as Napoleon Chagnon and Richard Wrangham, who have depicted them respectively as "fierce people" and "demonic males."[6] Science still views violence and warfare as integral to our species heritage, even though evidence for such behavior during our prehistory is slim. The archaeological record, for example, contains zero evidence for large-scale killings before the agricultural revolution of twelve thousand years ago. This makes war-is-in-our-DNA evolutionary scenarios highly speculative.[7]

The third reason we hear little about female leaders in other primates is perhaps the most significant. We often reduce social dominance in other species to physical dominance. How could it be otherwise? You either dominate or you don't. If Mama can't physically beat up any adult males, why even call her the alpha female? That we apply this simplistic logic to animals is surprising, given that we never do so in our society. No one visiting a business walks up to the most strapping guy in the office, convinced that he must be the boss.

The same holds for other primates. The biggest and strongest male isn't necessarily on top, because networking, personality, age, strategic skills, and family connections all help individuals ascend the social ladder. Applied to the genders, this means that a bonobo female may rank above everyone else in her community despite the presence of males far more muscular than herself. Among chimpanzees, even the smallest male may become the alpha. To do so, he needs the support of others. This introduces complexities: he needs to keep his allies happy, making sure they don't conspire with his rivals, and to win over females by protecting them and generously sharing food. Field studies show that the smaller a chimpanzee alpha male is, the more time he devotes to grooming others.[8]

Even the strict hierarchy of monkeys is less straightforward than you might think. Remember how Mr. Spickles, the aging boss of a rhesus monkey troop, relied on the backing of the alpha female Orange?[9] This makes you wonder who is the more powerful of the two. Already in the early years of our discipline, the father of Japanese primatology, Kinji Imanishi, observed that "although the society of monkeys may appear to be under the dictatorship of a mighty male monkey, females actually have great influence in the society."[10]

So let's pick apart social dominance. It has three components: fighting ability, formal rank, and power. Since young primates play-wrestle all the time, they quickly learn who is stronger or weaker. They simply feel it when they try to hold someone down or escape their grip. Like us, they become connoisseurs of physical strength by just looking at one another's build and gait. A female bonobo is keenly aware that she needs a sisterly alliance to dominate males. Males, too, know exactly where they stand physically, but since they often rely on alliances, body mass is a poor predictor of their hierarchy.

Female apes rarely compete physically with each other over status. In captive settings, we sometimes put them together from a variety of sources. It is astonishing how quickly females establish ranks. One of them walks up to another, who submits by bowing, pant-grunting, or moving out of the way. That's all there is to it. From then on, the first female dominates the second.

The contrast with males is dramatic. The many chimpanzee male introductions I have witnessed are always tense. One of the males may try to intimidate the other, which may provoke a fight, or else the two of them delay a confrontation for a couple of days, sometimes weeks. At some point, however, there's always a test of strength. This is why the most vigorous males, around the age of twenty, initially take the top spots. Once males get to know each other, however, they begin to form political alliances that rearrange the hierarchy. This is when smaller and older males come into play and rise in status.

In the age-graded system of females, being older is a good thing. Competition over rank is uncommon because high status doesn't do much for females in the forest, where they often travel and forage on their own. It's not worth the trouble males go through. One of the senior females is usually the alpha despite the presence of females in their prime, who would have no trouble winning a physical fight.

We know about female physical strength from handgrip tests that we conducted with our chimpanzees. In contrast to women, whose handgrip strength begins to weaken only in their sixties, in female chimpanzees it drops off already after their mid-thirties.[11] At that age, females become increasingly frail, yet they have no trouble holding on to their place on the social ladder. On the contrary, they often gain in status. Mama, for example, remained alpha until the day she died, at fifty-nine. She was nearly blind and walked unsteadily, yet she still enjoyed plenty of respect. Had Mama been a male, she'd have lost her position years before. In the wild, too, female chimpanzees achieve high status with age. They wait their turn for this moment in the sun, a process that has been described as "queuing."[12]

ONCE A HIERARCHY has been formed, it has to be communicated. Every social mammal has its submissive rituals, from a dog rolling onto his back with tail between his legs to a macaque baring her teeth in a wide grin. Chimpanzees and bonobos utter distinctive repetitive grunts while bowing for the dominant. An alpha male chimpanzee only needs to walk around with his hair slightly on end for everyone to rush forward and grovel in the dust while uttering panting grunts. Alphas underline their position by moving an arm over the others, jumping over them, or ignoring their greeting as if they don't care. Mama received far fewer

A subordinate male chimpanzee (left) bows, bobs, and pant-grunts for the dominant. The size contrast created by this status ritual is artificial. In reality, both males are the same weight.

submissive gestures than the males, but since every female in the colony showed them to her, and she never to them, she counted as the top female. These outward signs of status express the *formal* hierarchy in the same way that the stripes on military uniforms tell us who officially ranks above whom.[13]

Power is something else entirely. It's the influence an individual exerts on group processes. Like a second layer, power hides behind the formal order. Social outcomes in a chimpanzee group often depend on who is most central in the network of social ties and alliances. When Mama decided that the colony's hostilities against young alpha male Nikkie had to come to an end, she showed herself to be more powerful than him.

Nevertheless, Nikkie was the formal leader, receiving submission from all other members of the community. A few months before the incident, he had defeated the previous alpha male with the help of his older buddy, Yeroen. By hooting along with Nikkie and backing him in his campaign, Yeroen had created a situation that worked out very well for him. As the boss, Nikkie was forced to treat Yeroen with kid gloves and let him mate with females. The old male didn't have the vigor and stamina to be alpha himself, but he had regained power and respect as kingmaker.

These configurations are also known from the wild. I was pleased to meet Kalunde in the Mahale Mountains, in Tanzania, after hearing so much about him from Nishida. In real life, Kalunde was smaller than I had expected. He had "shrunk" in his old age, Nishida explained. Kalunde had moved into a key position in the chimpanzee community by playing off younger males against each other. These ambitious males had sought Kalunde's support, which he handed out erratically, making himself indispensable to any one of them. Being a former alpha male, Kalunde made a comeback of sorts, but like Yeroen, he did not claim the top spot. Instead, he acted as power behind the throne. Nishida and I compared notes at night in camp and were struck by the eerie similarity in tactics between Yeroen and Kalunde. Both were over the hill, like the elder statesmen in our capitals. But also like them, they were still very much in the thick of politics.[14]

What we have then are social hierarchies in which physical strength is a big plus for males, but not so much for females, for whom age and personality are more important. And even though rank is communicated through formal signals, it is not the best indicator of political power. The Burgers' Zoo colony, for example, was essentially run by Yeroen and Mama, the oldest male and female, despite both of them ranking below the top. With Yeroen having the young alpha male in his pocket and Mama's ability to mobilize all the females, no one could get around this duo.

In her heyday, Mama was an active player in male power struggles. She would rally female support for one male or another, and the one who managed to get to the top would be in her debt. And if Mama ever turned against him, his career might be finished. She acted like a party whip on behalf of her favorite male by punishing females who dared side with his rivals. She had an excellent memory for infractions. Once she waited until all the chimpanzees had entered the night building, then cornered and roughed up a female who had sided with the "wrong" male earlier in the day.

When we say that chimpanzees are male-dominated and bonobos female-dominated, we therefore need to qualify that the less dominant sex is never powerless. And apart from the three primary status markers—fighting ability, rank, and power—there is a fourth: *prestige*. Prestige is vital in species, such as ours, that rely on the transmission of knowledge. As cultural beings, our attention automatically goes to the most experienced and skilled among us. We look up to our heroes and emulate them. Teens try to dance like Beyoncé, and men want the same wristwatch as Roger Federer. Prestige is a form of power that comes from being admired.[15]

Since apes learn plenty of things from each other, too, we can expect the same tendency. In one study, we let chimpanzees watch two group members demonstrate behavior that we had taught them. One model was a female of high status, the other one of low status. In the presence of the entire colony, these females were rewarded every time they dropped a plastic token into a box. Each female had her own box with different markings. Even though both models were equally visible and successful, it was as if the chimpanzees had watched only one of them. Emulating the high-ranking female, the colony began to massively put tokens into her box while ignoring the other's box.[16]

Prestige isn't imposed from above but conferred from below, which makes it more sophisticated than the physical coercion everyone expects in a primate society. The same is true for the power structure, which only loosely corresponds with fighting abilities. Thus if anyone claims to know which gender is naturally dominant, we should always ask what they mean.

AMOS, AN ALPHA MALE chimpanzee at the Yerkes Field Station, was strikingly handsome and beloved by his group. The first observation reflects human judgment, which the apes may not have shared. The second, however, received irrefutable proof during Amos's dying days.

We learned postmortem that in addition to a massively enlarged liver, Amos had several cancerous growths. Even though his condition must have been building for years, he had acted normally until his body couldn't hold out any longer. Any hint of vulnerability might have meant loss of status, which is why males tend to hide weaknesses and act stoic around their rivals. We found Amos in his night cage panting at a rate of sixty breaths per minute, with sweat pouring from his face, while his fellow apes sat outside in the sun. Amos refused to go out, so we held him apart. Since the other chimpanzees kept returning indoors to check on him, we cracked open the door behind which Amos sat to permit contact.

Amos placed himself right next to the opening. A female friend, Daisy, gently took his head to groom the soft spot behind his ears. Then she started pushing large amounts of wood wool through the crack, which chimps love to build nests with. Amos was leaning against a wall without doing much with the wood wool. Daisy reached in several times to stuff the wool behind him between his back and the wall. It looked exactly like the way we arrange pillows behind a patient in the hospital. Other chimpanzees brought wood wool also.

The next day we put Amos to sleep. There was no hope for survival, only the certainty of more pain. Some of us cried upon his death, and his fellow apes were eerily silent for days. Their appetite took a hit. Amos had been one of the most popular alpha males I have ever known.

Amos's position defies the stream of modern business books, such as the *Alpha Male Bible* (2021), intended to instruct men how to become an alpha male. These books teach body language tricks and urge men to think like a winner with the goal of securing the corner office and charming women. They forget to mention the skills that set a good chimpanzee alpha male apart, such as generosity and impartiality. We're presented with a cardboard version of the alpha concept, which I find all the more galling given how much my book *Chimpanzee Politics* contributed to its popularity.[17]

I recognize two main alpha types. The first type fits the one fêted in these business books. They are bullies who live by Machiavelli's credo that "it is better to be feared than loved if you cannot be both." These males terrorize everyone and are obsessed with instilling loyalty and obedience. We know this kind of male all too well in our species but also in chimpanzees. They are disturbing to watch. In Gombe National Park, for example, Goblin was one of those exceptional alpha males who acted like a jerk from an early age onward. He'd kick other chimps out of their nests in the early morning without any reason at all. He was known for picking fights and never giving up, even against his former protector and mentor, the reigning alpha male. Eventually, he pushed this male off his throne. Rather than making friends, his favorite tactic was physical intimidation.

One day Goblin got his comeuppance, after he unexpectedly lost against a younger challenger. A mass of angry apes set upon him, as if they had been waiting for this opportunity. A screaming Goblin emerged from the huge brawl in the undergrowth. Running away, he had injuries on his wrist, feet, hands, and scrotum. He would almost certainly have died from infections had he not been darted by a veterinarian who gave him antibiotics.[18]

The other type of alpha is a true leader. While he is dominant and defends his position against rivals, he is neither abusive nor overly aggressive. He protects the underdog, keeps peace in the community, and reassures those in pain or distress. Analyzing all instances in which one individual hugs another who has lost a fight, we found that females generally console others more often than males. The only striking exception is the alpha male. This male acts as the healer-in-chief, comforting others in agony more than anyone else. As soon as a fight erupts, everyone turns to him to see how he will handle it. He is the final arbiter of disputes.[19]

For example, a quarrel between two females spins out of control and ends with hair flying. Numerous apes rush up to join the fray. A knot of fighting, screaming apes rolls around, until the alpha male leaps in and beats them apart. He does not choose sides, unlike everybody else. Instead, anyone who continues to fight will receive a blow. Or he strides in between two screaming parties and stands there imposingly with all his hair on end, making it clear that they will have to push him out of the way to continue. Sometimes an alpha male will raise both arms as if imploring contestants to stop the fracas.

Known as the *control role,* this constructive attitude isn't found in all primates. It's totally absent in baboons, for example. In the Serengeti, in Kenya, the American neuroscientist and primatologist Robert Sapolsky studied anxiety in baboons by measuring stress hormones in their blood. Older males have mellowed, but young adult ones are engaged in a relentless war of nerves. They freak everybody out with their long, pointy canine teeth. Sapolsky left no doubt that the male hierarchy is full of nastiness, fear, and random violence. The alpha male himself doesn't escape stress either, as he has to look over his shoulder all day, because others are eager to steal his position. There's no sign of him coming up for the underdog or fostering social harmony.[20]

Keeping order and breaking up fights is typical, however, of primates dominated by a single large male, such as gorillas and hamadryas baboons. The male regularly steps in to restore the peace among females.[21] Male chimpanzees go further in that they control a much wider range of internal disputes. Initially, this behavior was best documented in zoo settings, but then the American anthropologist Christopher Boehm, after studying human societies, spent two years at Gombe National Park. He discovered that wild chimpanzees, too, head off fights between others. Given their flexible gatherings, the alpha male isn't always around, so the task falls to the highest-ranking male on the scene. In the below example, Gombe's beta male, Satan, broke up a confrontation between two adolescent males:

He charged right at the protagonists, but they were so involved in their conflict, grappling and trying to bite, that this approach had no effect. Satan, an unusually large male, first thrust juvenile Frodo aside because he was standing nearby and might have entered the conflict. Then he put his great arms between the bodies of the two combatants and literally pried them apart, which took him a full four seconds.[22]

Primates typically favor relatives, friends, and allies in everything they do, but the control role is different. Policing males place themselves *above* the fray. Their impartial intercessions aim at restoring the peace rather than aiding friends and relatives. And in case they do favor one party over another, their choice doesn't necessarily match their social preferences. They protect the weak against the strong, such as a female against a male or a juvenile against an adult. The male who adopts this role is the only even-handed member of society.[23]

A community doesn't automatically accept the authority of would-be arbiters. When Nikkie and Yeroen dominated the Burgers' colony as a team, Nikkie tried to step in when disputes arose. More often than not, however, he ended up on the receiving end. Older females did not accept this snotty-nosed upstart, whom they had known as a little juvenile, coming in and hitting them over the head. Nikkie was also far from impartial: he sided with his friends regardless of who had started the fight. Yeroen's attempts at pacification, in contrast, were always accepted. He did so with fairness and a minimum amount of force. In due time, the old male took over the role from his junior partner. Nikkie wouldn't even bother getting up if a fight erupted, leaving its settlement to his senior partner.

The above shows that the second-in-command can perform the control role, and that the group has a say in who plays it. Everyone throws their weight behind the most effective arbiter. This gives him or her the broad authority to maintain law and order and shield the weak against the strong. I add *her* because if females got into a fight, Mama would not hesitate to perform the same role. She was so highly respected that this never posed a problem.

Mama and other females also sometimes "confiscated" the weapons of males. If two rivals were gearing up for combat—hooting, body-swaying, and gathering heavy rocks—a female might walk up to one of them to disarm him. The male wouldn't resist her removing a rock from his hand. Once the confrontation started, however, the situation became too risky for females to get involved. They'd only step in en masse on the rare occasions that things escalated to bloodshed.

How much a group benefits from mediation and arbitration was experimentally demonstrated in pigtail macaques. In this primate, too, high-ranking males police fights between others. Working with a troop of more than eighty monkeys in a large outdoor enclosure at the Yerkes Field Station, my graduate student Jessica Flack and I kept three top males out of the group. We did so only for one day at a time. During those days, the society seemed to come apart at its seams. There was less play, and the monkeys fought more. Fights lasted longer than usual and turned violent more often. With the leading males absent, reconciliation rarely followed these skirmishes. As a result, tensions among the monkeys rose to a worrisome level. The only way to restore stability was to return these males to the troop.[24]

This experiment demonstrated the degree to which dominant individuals contribute to social harmony. They are essential to keeping a troop together.

TROPHY HUNTERS, by eliminating the most magnificent specimens of a species, enact reverse selection. It's the opposite of natural selection. The hunters remove the healthiest and fittest males from the gene pool by targeting the largest bears or the lions with the darkest manes. The same sort of reverse selection has had disastrous consequences for elephants, in which it combines with ivory poaching. In many populations, bulls with large tusks have gone virtually extinct. One of the devastating side-effects has been that young bulls have become unruly and dangerous.

In Pilanesberg National Park in South Africa, marauding gangs of juvenile elephant bulls went berserk. Like a blood sport, they began to chase down white rhinoceroses, stomping them with their feet and goring them to death with their tusks. They harassed other animals as well. The park resolved this problem by setting up a Big Brother program. Park staff flew in six full-grown bull elephants from Kruger National Park. Bulls keep growing larger throughout their lives, and the oldest ones often roam with younger bulls in tow. Like warriors in training, the latter follow and watch their mentors. The hyperaggressive state of *musth*—when testosterone levels increase fifty-fold—is curbed when young bulls are exposed to dominant males. A young bull may lose the physical signs of *musth* within minutes of being put in his place by a bigger one. At Pilanesberg, hormonal suppression and reduced risk-taking in the presence of intimidating adults made all the difference. After the Big Brother program, signs of random violence disappeared. In previous years, elephants had killed over forty endangered white rhinos. The civilizing influence of older bulls stopped the carnage.[25]

In chimpanzee society, too, adult males serve a socializing function. Zoos have learned that to make a good alpha, a male needs to be of a certain age and background. Males who have barely left puberty behind or have grown up in the absence of older males often fail to bring peace and harmony. They are so volatile that they stress everyone out. The discipline and mentorship provided by adult male models is essential for young males to mature into emotionally stable characters.

Discipline by adult males is especially striking since young chimpanzees go unpunished during the first four years of life. They can do nothing wrong, such as using the back of a dominant male as a trampoline, ripping food out of the hands of adults, or hitting other juveniles as hard as they can. Every conflict is quickly soothed, and whenever juveniles are about to commit a faux pas, elders distract them. One can imagine their shock and panic when, after years of enjoying this puppy license, they are punished for the first time.

Adult males mete out the severest reprimands to young males who fail to show appropriate submission, who bother females and their young, or who try out their nonexistent sexual skills on fertile females. Most of the time the adult male merely chases or hits the youngster, but sometimes he inflicts injuries. A young male needs only one or two harsh lessons to get the message. From then on, every adult male can make him jump away from a female by a mere glance or a step forward. It is all part of an extended education in impulse control. Young males learn boundaries, become circumspect before they act, and keep an eye on dominant males. They also follow their elders around and mimic their behavior. If a colony's alpha male, for example, combines his bluff displays with spectacular jumps, you can be sure that all the young males will soon be performing similar jumps. In the natural habitat, too, younger male chimpanzees seek out older ones as role models.[26]

When the above elephant story hit the media, commentators made the inevitable connection with human families. About a quarter of children in the United States grow up in fatherless households. Children in such families suffer from more behavioral problems, substance abuse, school failure, and suicide. In accordance with the idea that boys externalize problems and girls internalize them, sons in mother-only families direct their anger outward and often turn violent or delinquent. In contrast, daughters suffer from low self-esteem and depression and are at greater risk of teenage pregnancy. Cause and effect are notoriously hard to pinpoint in human social studies, but data do suggest that having both parents in a family has a stabilizing effect.[27] Since children need same-gender role models, the presence of a father figure is beneficial even if not absolutely critical. Lesbian parents, for example, often bring male figures into their children's lives. They invite stand-in dads into their homes or encourage children to seek interaction with uncles, male teachers, or male coaches.[28]

It was long thought that father absence mainly affects household income and by extension family stress levels, but we can't exclude that it also has a hormonal impact. Like older elephant bulls who inhibit the *musth* of younger ones, hormonal suppression is also known in primates. For example, young male orangutans don't develop secondary sexual characteristics (such as fleshy cheekpads, or flanges) as long as a large male is around. Their growth is stunted until the day the old male dies or is moved out. In a Sumatran forest, the downfall of a resident male immediately sent two adolescent males into a growth spurt. At zoos, the same effect is well known and sometimes may be attributable even to human males. One orangutan male refused to grow up, one story goes, staying lanky and immature-looking for years. The zoo veterinarian couldn't find anything wrong with his health. One day the longtime primate keeper took his retirement. Within a few months, the young orangutan developed full-size flanges and a luxurious coat of orange hair. Apparently, the mere aura of this man had stifled the ape's growth.[29]

In our species, too, having a man in the family may affect children's hormone levels. Fatherlessness seems to accelerate puberty. One study asked more than three thousand American women and men about the age at which they experienced the first signs of menarche (women) or lowered voice (men). It was found that growing up without a father brought an earlier puberty onset for both genders. The absence of a mother didn't have the same effect. After a separation or divorce, many things may change in a family, including a loss of income or a move to another neighborhood, which makes it hard to tell what exactly causes the difference. Possibly, though, the daily presence of a father slows down a child's hormonal development.[30]

The above makes clear that the role of dominant male primates shouldn't be dismissed or cast in a negative light, as if every one of them were a tyrant. There are indeed males who terrify everyone, but they aren't the rule. In our closest relatives, the majority of alpha males that I have known didn't harass or abuse the members of their society. They guaranteed peace and harmony by keeping order and checking the behavior of up-and-coming young males. The security that an even-handed alpha male provides, especially to the most vulnerable, may make him immensely popular. He is massively supported whenever a challenger emerges. And when he loses his position, which inevitably happens one day, he simply steps down a few rungs on the social ladder and lives out his life in peace.

Like Amos, he may enjoy love and affection in his final days.

EVERY PRIMATE GROUP has one alpha male and one alpha female and not an alpha individual (either sex) followed by a beta individual (either sex), then a gamma, a delta, and so on. The reason is simple. Hierarchies are largely sex-segregated. In the same way that young primates and children prefer to play with members of their own sex, social hierarchies mostly involve either one or the other sex.

Females worry about where they rank relative to other females, and males do the same relative to other males. Competition occurs primarily within each sex, and hierarchies help regulate and contain it. Males compete with each other over status and who will mate with females. For females, in contrast, sex is less important than food. In evolutionary terms, a female's key to success is nutrition. She needs access to prime feeding spots to grow a fetus, nurse her infant, and feed her young. Since a female ape's offspring stays with her for a minimum of ten years, she faces much higher food demands than a male.

There is little reason for competition between the sexes. Both a male chimpanzee (who ranks above the females) and a male bonobo (who ranks below them) keep their eyes mostly on fellow males and invest aggressive energy into climbing the social ladder among them. For a female, too, it's all about maintaining her standing among other females. Ranking above or below the males is of less concern for her, as she travels, forages, and socializes most of the time with members of her sex. Males and females dwell in different worlds, each with its own set of issues.

Since science has traditionally focused more on the male than on the female world, we know remarkably little about the leadership styles of alpha females. I have already described how Mama at Burgers' Zoo and Princess Mimi at Lola ya Bonobo were politically astute and firmly in control. Mama had her faithful female ally, Kuif, who stood by her through thick and thin. And Mimi, like every bonobo alpha female, relied on a powerful clique of central females. Mama's forte was to fix things after confrontations within the colony. Whereas high-ranking males control fights by stepping in while they are ongoing, Mama went into action afterward by smoothing things over and bringing the parties together.

If two male rivals failed to reconcile, for example, they'd often hang around each other. During their deadlock, they'd carefully avoid making eye contact, like two angry men at a bar. Mama might walk up to one of them and start grooming him. After several minutes, she'd slowly walk toward the other male, often followed by her grooming partner. If he failed to come with her, she might turn around and tug at his arm to make him comply. After the three individuals gathered for a while, with Mama in the middle, she'd get up and wait nearby until both males were grooming each other.

I have seen other chimpanzee females—always older ones with high authority—accomplish similar missions. For example, Ericka, the alpha female of Amos's group, was known among us as the "grooming machine." Always busy grooming everyone, she was so popular that others lined up to receive her attentions. She would do grooming rounds especially after scuffles. Since primates tend to synchronize their activities, Ericka's grooming was contagious, and others would follow her lead. By creating large clusters of grooming apes, she'd get the whole group to calm down.

Alpha females in the wild do not always take such a central position. In the best-studied communities, which are mostly in East Africa, chimpanzees live dispersed over the forest. Here females tend to stay out of the fray. They lack the proximity of other females to protect them if the males get rowdy. Females often carry their youngest offspring on their belly or back, which makes them vulnerable. They avoid taking unnecessary risks. In West Africa, on the other hand, chimpanzee females often travel together. Their intense social life is more like that in captive colonies. Females show solidarity and maintain lifelong friendships, supporting each other at every turn. Top-ranking females carry more weight in these wild communities and aren't reluctant to get involved in power politics.

In Taï Forest in Ivory Coast, Christophe Boesch described how some females would push and elbow themselves into a meat-sharing circle and take as good a place at the table as the males. These females made sure the alpha male had meat. If a scuffle broke out about possession, they'd support this male so that he could grab a piece. Since he shared with them generously, it was a great deal for both. Female friendships in Taï lasted for years, perhaps a lifetime. If a female's best friend went missing, she would search for her, whimpering in distress. Their loyalty to each other extended to offspring. In the same way that Mama adopted the youngest daughter of her best friend Kuif after the latter's death, females at Taï have been known to care for a deceased friend's dependent son or daughter.[31]

Female apes make sure their progeny are protected and well-fed, but having only one offspring at a time limits how many they can raise.

Another way they can grow their dynasty is through their sons. Daughters leave the community at puberty, but sons stay. While chimpanzee mothers sometimes help their sons advance in the male hierarchy, the champions in this regard are bonobos. Here female friendships and solidarity are even more reliable, and mothers make formidable allies. The worst fights in a bonobo community occur when females get involved in male status struggles. Kame, the alpha female in the Wamba forest in the DRC, had no fewer than three grown sons, the oldest of whom was the alpha male. When old age began to weaken Kame, she became hesitant to defend her children. The son of the beta female sensed this because he began to challenge Kame's sons. His mother backed him up and, in typical bonobo fashion, had no fear of attacking the top male on his behalf. The frictions escalated to the point that both mothers exchanged blows and rolled over the ground. Kame was held down. She never recovered from this humiliation. Her sons dropped in rank and after Kame's death became peripheral.[32]

Paternity data show that bonobo sons whose mothers are alive and well are three times more likely to sire offspring than are sons whose mothers died before they reached adulthood. Mothers actively interfere in sexual affairs by protecting their sons' courtship and helping them chase off rivals. Martin Surbeck, a Swiss primatologist, describes one such incident in the LuiKotale forest in the DRC:

> *Two of them—Uma, a female, and Apollo, a young, low-ranking male— were trying to have sex. Camillo, the highest-ranked male in the group, caught wind of their liaison and tried to come between them. But Hanna, Apollo's mother, rushed in and furiously chased Camillo away, allowing her son and his mate to copulate in peace.*[33]

The closest human parallel to females promoting their sons' reproduction is the fierce competition and intrigues among enslaved concubines in the Ottoman imperial harem. Some of these women gained status equal to that of the sultan's wives. If one of them bore a son, she'd be sent off to raise him and have no further children. Mothers lobbied intensely for their son to become the next sultan. The winning son, upon his ascension to the throne, would order the killing of all his half-brothers. His fratricide guaranteed that he'd be the only one siring offspring.[34]

We humans do things more rigorously than bonobos.

The connection between social status and reproduction has been lost in modern society, thanks to our prosperity and our access to effective birth control. Human psychology, however, can't shake the effects of this age-old connection. Since our inborn tendencies derive from ancestors who spread their genes, their means of social success are engraved into our psychology. Both male and female primates, and both men and women, are eager to ascend the social ladder. This has always been the winning ticket.

Our primate heritage is still visible in the way we evaluate male and female leaders. We pay attention to the physical size of men, for example, but not of women. You'd think that we'd pay at least as much attention to a man's intellect, experience, and expertise, but we remain stubbornly sensitive to his height. Our biases echo a time when physical prowess mattered more.

Height is positively associated with salary and even plays out in political debates. In forty-three U.S. presidential elections between 1824 and 1992, the taller candidate had double the chance of becoming president. This advantage explains why vertically challenged politicians, such as Prime Minister Silvio Berlusconi of Italy and President Nicolas Sarkozy of France, used to travel with a box to stand on during photo opportunities. Sarkozy sported stack heels whenever his tall model wife accompanied him.

A study by Mark van Vugt, a psychologist at the University of Amsterdam, presented subjects with photographs of men and women wearing business attire. In some photos the candidates looked taller, and in others they looked shorter, due to manipulated background elements. The subjects preferred tall men as leaders ("This person looks like a leader") based on their perceptions of dominance and intelligence. Regarding women, on the other hand, height had minimal effect.[35]

If being tall enhances a man's status prospects, does age do the same for women? This would fit with what we know about our primate relatives. The world has seen many postmenopausal women as heads of state, such as Golda Meir, Indira Gandhi, Margaret Thatcher, and German chancellor Angela Merkel, the most powerful woman of our own era. In more recent times, however, younger female leaders have emerged as well. Some first-rate leaders, such as Prime Minister Jacinda Ardern of New Zealand, are in their childbearing years.

It has been suggested that female leaders have done particularly well during the COVID-19 pandemic. The data aren't final, though, and it is hard to compare countries due to confounding variables such as population size, health care system, and GDP. But we can at least say that quite a few prominent male leaders have failed miserably. As Nicholas Kristof put it in *The New York Times,* "It's not that the leaders who best managed the virus were all women. But those who bungled the response were *all* men, and mostly a particular type: authoritarian, vainglorious, and blustering."[36] One theory is that women leaders feel less pressure to appear strong and decisive. They have enough humility to call in experts and follow their advice. They also seem to sympathize more with those who arc affected and to appeal to the public to contain the threat. By contrast, some male leaders have treated the virus almost like a personal affront and made attempts to dominate it by means of political rhetoric rather than medically proven countermeasures.

Perhaps male and female leadership comes with different strengths and weaknesses. Based on our primate background, we would expect males to be good at impartial arbitration. There are two reasons why male primates more often wade into an altercation to put a halt to it. First, their more intimidating physical presence draws immediate attention and sends a warning to anyone who wishes to continue the brawl.

And second, it's easier to be impartial if one doesn't need to take close relatives into account. Males may have offspring in the group, but their knowledge of paternity is vague or absent. On the other hand, females often have offspring and grand-offspring, sometimes a dozen of them, all of whom they know individually. Given how fiercely females defend their kin, it is nearly impossible for them to stay neutral during confrontations within the group.

This doesn't mean that females are incapable of performing the control role. I learned this one day at the Wisconsin Primate Center in Madison, when a colleague, Viktor Reinhardt, drew my attention to a troop of rhesus monkeys dominated by an older female named Margo. What struck us about her was not so much her rank as her ability to keep the peace. Margo was large, but not exceptionally so. I was used to the alpha female Orange, who was the matriarch of such an extended matriline that it was entirely understandable that she left the policing to Mr. Spickles. He did a first-rate job maintaining order while Orange kept her daughters and granddaughters in the upper echelons of the female hierarchy, as any macaque matriarch does. But Margo was different. Not only did she rank above everyone in her troop but she was childless.

Viktor studied this troop and concluded that even though every other monkey interfered in fights to help friends and kin, Margo showed no such bias. She was as good at the control role as was Spickles, and since she had no family to worry about, she could be just as even-handed. She systematically defended the downtrodden. She would go out of her way, occasionally performing fierce attacks, to protect bottom individuals regardless of their sex or age. They'd crouch in front of her in fear of their aggressors, and Margo would rest her hand on them to leave no doubt whose side she was on. Just like males performing the control role, Margo seemed to act out of community concern.[37]

These observations suggest that the typical behavior of the sexes doesn't tell us everything about their capacities. Each has *potentials* that manifest themselves under rare circumstances. Female primates may possess an excellent potential to perform the control role if they're freed of kinship obligations. This potential is relevant to the modern human workplace, where bosses rarely have to deal with kinship relations. In fact, we wisely have rules against nepotism at work, which are designed to keep family bonds out of the picture.

In contrast to primate social life, modern society is large in scale, and we tend to integrate both sexes into a single framework. This is a strikingly new development in our evolutionary and cultural history. Anthropologists call tribal societies "hunter-gatherers" precisely because women go out in groups to gather fruits, nuts, and vegetables, while the men form hunting parties. The women chat, gossip, and sing together while they are away from camp, whereas the men often walk long hours in silence so as not to attract attention. These roles were perhaps never as separate as is often assumed (we know of women hunters and warriors), but throughout most of our history and prehistory, labor was divided by gender. Women shrugged their shoulders at the typical activities of men, and men at those of women, even though both depended on each other. Only in the industrialized era have we begun to mix things up by weaving both spheres together. Corporate jobs require men to take orders from women and women from men. We ask both genders to respect and rely on each other's work.

Since men initially designed the corporate environment for themselves, gender debates often focus on how to make it more welcoming for women. There is, for example, the pervasive myth that men are more hierarchical than women. Doesn't this make the workplace, with its layered social organization, a hostile place for women?

Insofar as this argument assumes women to be nonhierarchical, I have great trouble with it. In almost every social animal, both sexes arrange themselves on a vertical scale. After all, the very term *pecking order* comes from hens, not cocks. Everyone who has watched female baboons, or female bonobos for that matter, will be quickly disabused of notions about female egalitarianism. The same applies when women

A woman curtsies for the alpha female, a queen. Whereas
we tend to associate hierarchies more with men than with
women, they mark both genders.

spend much time together, as in all-girl schools, female prisons, and
feminist organizations. I happen to be familiar with nuns and can't say
that Mother Superior struck me as antiauthoritarian. There is in fact no
data demonstrating that men are more hierarchical than women. The
only difference one study reported is that when people are put together
in same-gender groups, men settle on a rank order more quickly than
women. Women do eventually always form one, however.[38]

Even the small-scale societies that anthropologists call "egalitarian"
have to work hard to stay that way. They are by no means free of over-
bearing individuals. Other group members employ ridicule and gossip,
and sometimes harsher methods, to bring ambitious individuals down
to earth. That they need to use these countermeasures attests to the per-
vasiveness of our species' hierarchical inclinations. Whenever we seek to

accomplish something together, whether on a school board, in a garden club, or in an academic department, a pecking order—however vaguely defined—emerges. Harold Leavitt, an American psychologist of management, compared maligned corporate hierarchies to dinosaurs that refuse to die: "The intensity with which we struggle against hierarchies only serves to highlight their durability. Even today, just about every large organization remains hierarchical."[39]

Modern society's attempt to integrate both genders into a single hierarchy relies on the leadership capacities of both. From looking at other primates, we know that these capacities can be found in both sexes. They may not be exactly the same, but they overlap more than they diverge. We have no reason to assume, as is often done, that males are more suited for leadership than women. Men's greater size and strength doesn't make them better leaders, even though these qualities still subconsciously bias our judgment. In other primates, both sexes astutely exert power, and female leadership is not hard to find. Females also have a hand in the hierarchy among males in the same way that males have one in the hierarchy among females. Moreover, many alpha individuals, regardless of their sex, care about more than rank. They defend the underdog, settle disputes, console distressed parties, facilitate reconciliations, and promote stability. They serve their community while at the same time safeguarding their position and privileges.[40]

Instead of making Machiavelli's choice between love and fear, most alphas instill both.

10

KEEPING THE PEACE

Same-Sex Rivalry, Friendship, and Cooperation

When yard workers—always men—arrive at our home to mow the lawn or do some landscaping, they address me instead of my wife, Catherine, even though both of us are standing side by side right in front of them. They feel more comfortable talking to me. They expect me to tell them what to do, not realizing that the garden is my wife's baby. She knows every little corner, while I am about as ornamental as an azalea bush. It doesn't take them long to figure out who's boss.

Catherine will make an eyeroll. Men ignore women in politics, at car dealerships, in hardware stores, and many other places. There are several explanations, including, of course, plain old misogyny and disrespect. Many men just can't imagine that women know anything about what they consider male jobs. But the problem goes deeper. Not all men are misogynous, and not all of them automatically dismiss a woman's expertise. Men's selective attention often has less to do with women than with the presence of other men. We need to go to a more fundamental level to understand this response.

The reason humans take only a second to detect a person's sex is that throughout our evolutionary history this information has been crucial. Like all animals, we have different social and sexual agendas with our own sex than with the other one. We also have different fears. Thus a woman walking alone at night needs to quickly determine if a group of strangers on her path is all-male or mixed gender. The latter is much less worrisome.[1]

Our gender radar is always turned on. It's an illusion to think that we adapt to modern society without this evolutionary baggage. Our social software was written millions of years ago. For men, this means keeping an eye on their fellows. Since male-to-male combat has always been part of primate history, including ours, we can't expect men to turn off their selective attention. This applies even to environments high on trust and low on violence. At the office or the university, there is no shortage of intrigues and power plays. I have witnessed cursing, yelling, door slamming, coups, and betrayals. Such tactics are not limited to men, of course, but it's among them that verbal arguments most easily escalate to pushing, shoving, or other physical contact.

A curious case that warmed this old ethologist's heart was that of a mathematics professor at a California university who was accused of peeing against a colleague's office door. The two male professors were said to have had a dispute that escalated to a "pissing contest." After someone had found puddles in the hallway, school officials set up a camera and captured the urinating professor on video.[2]

To prepare for physical combat is an unconscious survival mechanism. It absorbs male attention not just for negative reasons, having to do with risk, but also for positive ones, because the best way to avoid conflict is to get along and make friends. I'll call it the *male matrix:* males are part of an exclusive network that makes them automatically tune in to members of their own sex. I here use the term *matrix* the way we do in biology, to refer to a connective tissue within which items (such as cells) are embedded.

The selective attention men pay to other men insults women since it cuts them out. They feel ignored. I am not defending this attitude, but I also feel that its existence shouldn't keep us from understanding where it comes from and how it compares to behavior in other primates. One can study a phenomenon without sanctioning it. The male matrix is moreover not without its female counterpart. Even if women don't seek to test their physical strength against each other, they compare other aspects of their physique. Rivalry is by no means bound by gender, and women keep an eye on each other as well.

In all primates, males compete overwhelmingly with males and females with females. This applies equally to us. When college students were asked to keep track of their daily competitive thoughts and encounters, the genders came out with similar results. Men and women compete to an equal degree among themselves and about the same things, such as academic performance and getting what they want. But college-age women envy their peers more for their looks, while their male counterparts compare themselves more on athletic prowess. Both qualities figure in mate competition, which peaks around this age.[3]

As illustrated by the scent-marking college professor, rivalry within each gender even steers chemosensory communication. We are mostly unaware of it, but one exceptional businessman believed in it. While I was sitting next to him on a long plane ride to Tokyo, I asked him why he went in person to his meeting. Couldn't he do it virtually? He laughed, saying that he wants to smell the other side. "I like to be in the same room," he said, "to see them sweat, to get their scent, and see their faces up close."

We are sensitive to the odors of others and actively try to catch a whiff. By surreptitiously filming people after a handshake, scientists discovered that the hand that has done the shaking often finds its way to the owner's nose. They measured how many seconds the hand spent there and even assessed the nasal airflow of some subjects. They found that after interactions within their gender, people take a moment to sniff their hand. Men and women do so equally: men with men and women with women. Handshakes between the genders don't prompt the same inspection. Automatic-looking gestures (arranging one's hair, scratching one's chin) bring the hand close to the face, offering traces of the other's aroma. Olfactory sampling allows people to assess the level of self-confidence or hostility of potential rivals. Even though humans use the opportunity to smell each other as predictably as rats and dogs, they do so largely unconsciously.[4]

Despite fundamental gender similarities in competition, psychologists have a habit of downplaying it for women while inflating it for men. Men are said to engage in relentless one-upmanship, while women are seen as empathetic and mutually supportive. Psychology textbooks still call women the more communal sex and juxtapose feminine sociability and desire for intimate connections with masculine hierarchies and the seeking of distance and autonomy. Psychologists marvel at the depth of female friendships and almost pity those of men. In her book *Friendship,* Lydia Denworth summarized, "Recent decades have brought a strong view that women excel at friendship and men are duds."[5]

That serious scientists fall for this contrast is exasperating given the contradictory evidence right under their noses. Every day we can see boys and men hang out together, do things jointly, play games, help each other, and chuckle at each other's jokes. Men immensely enjoy their gender's company. How could this be if all they get out of it is stress and competition? Little boys report the same satisfaction with their playmates as do little girls, and men enjoy lifelong friendships just as women do.[6] The reason we so often mention the "old boys' network" in corporate life and politics is that men love to do their buddies favors. They believe in reciprocity.

Women seek more intimacy and exchange of information with their friends than do men, who are more action-oriented and less inclined to disclose personal details. For this reason, the friendships of women have been called face-to-face and those of men side-by-side. Men like to do things together and often gather in a larger framework, such as a group of buddies. Both genders take pleasure in their own gender's company, and neither gender would like its friendships to be more like the other's. Women friends don't desire more shared exploits, while men friends aren't waiting for intimate revelations.[7]

So how did we end up with this bogus gender dichotomy? It has been around for decades despite the lack of support from observable behavior. Elevation of female sociality over that of males was a central theme of *Beyond Power,* a 1985 tome by the feminist author Marilyn French. Speaking of humanity's fictional prehistory before the patriarchy took over, French surmised: "The matricentric world was one of sharing of community bound by friendship and love, of emotional centeredness in home and people, all of which led to happiness."[8]

Reading this line, I had to think back to my brief membership of a women's rights group. It was an eye-opener for a young man who was naïve about the other gender. It taught me that women aren't always bound by friendship and love. They frequently undermine each other in a way that would be detailed in 2001 by the free-thinking American feminist Phyllis Chesler in *Women's Inhumanity to Women.* Chesler documented the gossip, envy, shaming, and ostracism to which women subject each other. This had escaped attention because women are taught to deny this side of themselves. In hundreds of interviews, Chesler found that most women recall having been victimized by other women but deny ever doing the same to others. This is logically impossible, of course.[9]

The situation is not unlike the egalitarian delusion that I noticed in the student protest movement of the 1960s. While this movement had a clear-cut hierarchy of leaders, followers, and minions and hence was anything but egalitarian, everyone cheerfully pretended that it was. Similarly, women may live a "good girl" delusion even though they can be quite mean to each other. We humans sometimes develop a curious amnesia about our own behavior.

Oddly enough, other academic disciplines view the genders in a diametrically opposite light. Traditionally, anthropology has depicted human society as a pact among men. Anthropology in previous centuries saw field reports from far-flung corners of the world about male bonding, men's houses, manhood initiation rites, brotherhoods, big-game hunting, and warfare. Women were mere property, perfect for marriage exchanges between neighboring tribes. As one critical paper put it, "anthropology has always involved men talking to men about men." The term *male bonding* was made famous by Lionel Tiger in his 1969 book *Men in Groups*. He saw male camaraderie as a propensity that evolved in the service of group defense and hunting. Still today, human society's cooperative and moral nature is often attributed to the high level of male solidarity required for intergroup warfare.[10]

This perspective has problems, too. It underlines not merely male cooperation but also male rule. This might explain Tiger's unease with the political ascent of contemporary women: "It may constitute a revolutionary and perhaps hazardous social change with numerous latent consequences should women ever enter politics in great numbers."[11]

I don't share Tiger's worry and am glad that modern anthropology has left its androcentric focus behind. Like primatology, the discipline has seen a massive influx of women, who have changed its outlook. But anthropology was not wrong to emphasize the universality of male cooperation. It's a striking feature of our species that makes us stand out in the animal kingdom. It's common to see female animals forage together, jointly defend their young, and coordinate for other reasons. Think of a herd of elephants or a hunting pride of lionesses. Male cooperation, in contrast, is harder to come by. Males often stay away from each other and meet only for battle. There are a few notable exceptions, such as lions, dolphins, and chimpanzees, but the real champions are men. Men team up remarkably easily. They work together all the time to the point of putting their lives in the hands of their fellows during big game hunting and warfare. Male teamwork is a hallmark of human society.

There is no need, however, to emphasize one gender over the other when it comes to cooperation. A recent meta-analysis covering fifty years of research, hundreds of economic games, and thousands of human participants failed to find a substantial difference in cooperativeness between men and women.[12] It's no stretch to call all humans, regardless of gender, born team players. I therefore propose that we marry the anthropological view of brotherhood with the psychological view of sisterhood. Both are plain to see and powerful.

The confusion stems partly from men's reputation for being competitive and hierarchical. No one is denying these tendencies, but it's a mistake to think that they keep men from getting along. As if men's only choice is to either be rivals, continually jockeying for status, or friends, who love each other to death. The curious thing about men, however, is that they often are both. They smoothly go back and forth between the two. They can be simultaneously friends and rivals without losing any sleep over it. Moreover, instead of their hierarchies hampering cooperation, they facilitate it. I know this dynamic firsthand, having grown up in a family of six boys.

As a small illustration, a 2018 Netflix show featuring American-Canadian comedians Steve Martin and Martin Short starts out with both men sharing the stage while trash-talking each other. They heartily laugh off every creative insult flung their way while emphasizing that they have been friends for decades. Their banter endears them to us, because somehow a close nonsexual male friendship, or "bromance," is more believable if there's a little edge to it. Who can one needle if not one's friend? It may seem paradoxical, but men are perfectly comfortable being socially tight and assertive at the same time.[13]

It's a paradox that they share with male chimpanzees.

WATCHING MONKEYS, one can easily see the male matrix at work. A typical macaque troop has fewer adult males than females, with the former very much enjoying the latter's attention. Males appreciate the females' fastidious grooming, turning this way or that so that they can reach every little spot under their arms, between their legs, and especially places that are out of their own reach, such as shoulders and backs.

While basking in all this attention, they often get erections, which the females casually ignore.

As soon as one bit of tension appears in the air, however—a noisy fight, an alarm call—the males perk up and check on each other. Where is the alpha male? Where are their buddies? Each male wants a quick overview of guy affairs, both for his own safety and to assert his position if necessary. This is the male matrix in operation. At such moments, females vanish from view. Who was in the fight? Who gave the alarm call? Was it just a dumb juvenile who can't tell a vulture from an eagle, or did it come from one of their fellows? And why is one male missing? Did he perhaps sneak off with a female? Male monkeys take in all this information at a glance and calm down only once everything is clear. Then they enjoy female company again.

In chimpanzees, males hang out not only with females but preferably with each other. The male matrix is tighter because the risks are higher while the mutual bonds are stronger. It even affects their participation in the cognitive tests that we conduct at the Yerkes Field Station. I always joke that we have more data on females because the males have no time for us. They are too busy with power and sex. We summon each ape by name to enter a small building with unbreakable glass windows separating them from us. Here they will work on computer screens and show us their skills at sharing food or using tools. Participation is voluntary, but since the tests don't take long and we offer them goodies in an air-conditioned space, most chimps are eager to come in.

The only exception are the adult males, who don't like to leave their buddies behind. First off, if a female with swollen genitals is around, they know that other males will take advantage of their absence to mate. They want to foil this possibility at all costs. Second, even in the absence of sexual distractions, leaving their friends behind could have negative repercussions. The others will play and groom together, thus forging bonds that exclude the male who spends time with us. Male chimps want to be included in everything their buddies do. Any male who does enter our testing facility will continually peek underneath the doors to see what is going on outside, or he will hoot and bang doors to let everyone know he is still alive and well. This disturbs our tests to such a degree that we often end up releasing the male. He will run outside to give a spectacular bluff display, making sure that everyone knows he's back.

The male matrix is enhanced by *sexual dimorphism*, which is the difference in size and appearance between males and females. Male chimps are bigger, heavier, and hairier than females. *Piloerection,* or hair raising, is a language by itself that communicates tensions among males. One male will notice another doing something that goes against the established order, such as approaching food or a female or bothering an ally. He will put up all his hair and sway his upper body from side to side in a slow rhythm that draws attention to his broad shoulders. He may also stand up on two legs and pick up a stick to make his point. He gives off warning signals so that the other backs off and steps back from the brink. Most of the time it works, and he makes his intentions clear without any need for escalation.

Our species's dimorphism is of a similar magnitude. We too pay special attention to male shoulder width, which is why suits have shoulder pads. But in a bipedal species like ours, the main sex difference is height, which makes men stand out in a crowd. Being almost six foot four, I am taller than most American men and a foot taller than the average woman. My height colors my perception whenever I enter a gathering of people. My immediate attention is drawn to other men at eye level. In the same way that everyone probably finds it most comfortable to walk next to someone who makes strides of equal length and moves at similar speed, talking with people of one's own height is physically more comfortable. If size unconsciously biases contact preferences, it strengthens the male matrix only further.

Numerous studies demonstrate that people judge men but not women by height. This bias is by no means unique to our species, which is why male animals have special signals to communicate their mass and physical prowess. The hollow sound of a gorilla's chestbeat conveys the circumference of his torso. The breaching by a humpback whale indicates the amount of water displaced when he smacks back into the ocean. Bull elephants form "boy clubs" in which males build a hierarchy to mingle without too many confrontations. No one messes with the oldest, largest bulls, who walk with their heads held high and dominate the scene around fertile cows.[14]

Throughout the animal kingdom, males inflate their bodies by raising their shoulders, spreading their fins or wings, and puffing up their hair or feathers. Look at a standoff between tomcats in your backyard, with their slow-motion back-arching and body expansion without touching each other. The dominant male in a macaque troop has a cocky walk with his tail permanently in the air, which makes his rank visible. Males often show off weaponry, such as claws, antlers, and canine teeth. The males of our species are no exception. Even tilting his head up or down affects how dominant a man is perceived to be. Angry men hold up their fists while thrusting out their chests to show off their pectorals. In a common movie scene, a seated man is insulted by a standing one; then the first one rises and towers over the other to ask, "Did you call me an idiot?" This instantly turns the tables. We're all exquisitely aware of male body size, even those of us who are not into posturing and intimidation. Like many men, I am turned off by macho behavior, but that doesn't mean that I don't reckon with it. Every man learns how to counter, mitigate, or defuse it.[15]

Soon after their voices break, boys begin to build muscle power to the point that they barely realize what is happening to their bodies. It takes place so rapidly that they perform feats of strength that a few months earlier had been unthinkable. An amusing case involved a friend of mine at university who was taller than me. One day we were talking while walking into a classroom. Once we sat down together, we both stared with astonishment at the doorknob in my friend's hand. He was not in the habit of carrying one around. Looking back at the door through which we had entered, we noticed its knob was missing. He must have ripped it off without realizing. This is how boys become aware of their physical strength.

Constitutional body strength is a glaring exception to the general rule that gender differences are gradual and overlapping. According to one American report, more than two out of three men can lift 110 pounds directly off the ground, but only 1 percent of women can do the same. A German study measured the hand-grip strength of young people and found that 90 percent of the women fall short of 95 percent of the men. Could training explain this? Not really, because even top female athletes, who were considerably stronger than most other women, lagged behind the men. The strongest female athlete in this study reached the strength only of the average untrained male.[16]

Differences in strength are an essential part of male-male interactions, always playing in the background and sometimes in the foreground. Primate males deliberately perform muscle power feats, such as shaking a tree, throwing things around, or drumming loudly on a hollow tree, all to warn everyone of their vigor and strength. I once saw a wild alpha male chimpanzee perform an extraordinary display during which he dislodged large rocks and let them roll down a riverbed with lots of noise. He made it look effortless, but the rocks were so massive that other males could not match his behavior despite their efforts. I am sure they got the message.

The male matrix lasts into old age, when it changes in nature. Analyzing twenty years of data from Kibale National Park, in Uganda, Alexandra Rosati and co-workers discovered an old boys' network in wild chimpanzees. In their twilight years, when they are around forty, male chimpanzees limit themselves more and more to positive relationships that are free of tensions. They become increasingly selective about whom they groom with, focusing on just a handful of friends whom they prefer as much as they prefer them. They have lost interest in fake friends. Some of their remaining friends are their brothers, but most are unrelated.[17]

A similar selectivity occurs in our societies, in which older men spend more and more time with fewer friends. Their shrinking social circle has been attributed to the human sense of mortality. Realizing that their lives are coming to an end, men shift attention to their more meaning-ful contacts and don't waste time on negativity. As usual, however, this explanation overestimates the role of cognition in human affairs. We need to rethink it, now that the exact same tendency has been found in aging male apes. As far as we know, they are oblivious to the impending end of their lives.

My favorite explanation is that both men and male chimpanzees mel-low with age and lowered testosterone levels. When they are young and highly competitive, they form friendships for their political value. When they grow older, on the other hand, this value begins to take a back seat, and they don't judge companions by their utility anymore. Postprime males get together purely for fun and relaxation, a luxury younger males can only dream of.

O F T H E F O U R highly developed same-gender tendencies of our species—male bonding and competition and female bonding and competition—it is the last one about which we know the least. Female competition used to be minimized and denied to the point that the primatologist Hrdy complained that "competition between females is documented for every well-studied species of primate save one: our own."[18]

In nature, female competition over food is widespread for the sim-ple reason that a female can't raise offspring without sufficient nutri-tion. A second reason for female rivalry arrived when our male ancestors began contributing to the family. Once our lineage had pair-bonding and paternal care, females began to vie for the best partners on the mar-ket. As a result, jealousy and mate competition mark girls and women as much as they do boys and men, even if the genders wage these battles with different weapons.[19]

The delusion of the nice and pacific human female is falling apart. We now recognize, for example, that bullying among schoolchildren isn't merely a boys' problem. Counting fights in the schoolyard, a Finnish team led by the psychologist Kirsti Lagerspetz observed fewer incidents among girls than boys. But when she asked children at the end of the day about fights, she was in for a surprise. Both genders reported equal numbers. This means that most conflicts among girls remain invisible to the naked eye. Unlike boys' physical quarrels, girls show indirect aggression and manipulation, such as spreading false rumors or giving another the do-not-speak-to-me treatment. These tactics have only intensified with the arrival of digital media.[20]

In the last two decades, books have arrived with titles such as *Odd Girl Out: The Hidden Culture of Aggression in Girls* and *Queen Bees and Wannabes*. These books detail the silent treatment, snarky comments, and derogatory notes to which girls subject each other in their intense rivalry over friends and popularity.[21] The Canadian author Margaret Atwood novelistically contrasted the torments among girls with the more straightforward competition among boys. At one point, the principal character in *Cat's Eye* complains:

> *I considered telling my [older] brother, asking him for help. But tell what exactly? I have no black eyes, no bloody noses to report: Cordelia does nothing physical. If it was boys, chasing or teasing, he would know what to do, but I don't suffer from boys in this way. Against girls and their indirectness, their whisperings, he would be helpless.*[22]

My interest here is not so much in the prevalence of conflict among girls compared to boys as in the management of conflict. If both boys and girls spend most of their time with their gender and bond with them, they must both have effective ways of dealing with competition. Girls seem to be affected more deeply, though, because their discord lingers longer. Asked by Lagerspetz how long they might stay angry, boys thought in terms of hours, whereas girls believed that it could be for either one minute or for the rest of their life![23]

Boys are pack animals, stressing loyalty and solidarity, whereas girls form serial one-on-one friendships. These friendships are more intimate and personable than those among boys, but they are also more fragile. Studies have found that they generally don't last as long and their termination can be painful and acrimonious. Social exclusion is a typical girl tactic. Whereas boys quarrel all the time, this rarely disrupts their friendships or even their games. Boys enjoy debating the rules almost as much as the game itself. Among girls, in contrast, quarrels tend to end their games.[24]

It is hard to find information on how human adults keep rivalries from ruining their relationships. All we know is that women are deeply troubled by competition and find it hard to get over it. For example, after same-gender contests—whether on the tennis court or during games in the laboratory—women exchange fewer hugs and handshakes than men. Women distance themselves more from their adversaries. "Nothing personal" is a typical male remark after a match or harsh exchange.[25]

None of this makes women any less social or cooperative than men. It may just be that women strike a different balance between the benefits of close relationships and the cost of conflict. Men get less intimate with each other and therefore contain the damage in case conflict erupts, whereas the stakes are higher for women. Since my take on this contrast is inspired by studying how chimpanzees manage conflict, let me first explain these observations before returning to human gender differences.

AFTER THE BLOODBATH at Burgers' Zoo that I mentioned in the Introduction, I decided to dedicate my research to peacemaking. Having witnessed the tragic consequences of its failure, I wanted to learn more about reconciliation, a behavior that I had discovered a few years before.

Reconciliation is a counterintuitive phenomenon that reunites two parties who were antagonistic to each other. You'd expect chimpanzees to stay as far apart as they can, but they do the opposite. Former opponents actively seek each other out. One of the first of thousands of reconciliations that I have witnessed took me by surprise. Shortly after a

A conflict between a female chimpanzee (right) and the alpha male is reconciled with a kiss on the mouth.

confrontation, two male rivals walked upright on two legs toward each other. They were fully piloerect, which made them look larger than life. The two males locked eyes and looked so fierce that I was sure a revival of their hostilities was in the offing. But then they kissed and embraced and took their time to lick the wounds they had inflicted on each other.[26]

The definition of *reconciliation* (a friendly reunion between former opponents not long after a fight) is straightforward and easy to apply in the field. The emotions behind this behavior are hard to pinpoint, though. The least that occurs—and this is already remarkable enough— is that negative emotions, such as aggression and fear, are overcome to move to a positive interaction, such as a kiss. Chimpanzees undergo this reversal remarkably rapidly, as if they switched a control knob in their mind from hostile to friendly. Humans, too, are masters at turning this emotional knob. We do so every day in a conflict-prone environment where we need to accomplish goals together. We need to suppress our bad feelings or leave them behind. And whenever they do erupt, we need to fix things afterward. We experience the transition from hostility to normalization as forgiveness. This emotion is sometimes touted as uniquely human, religious even ("turn the other cheek"), but it may be natural to all social animals.

It took two decades before primatologists confirmed this phenomenon in wild chimpanzees. Although less common than in captivity, reconciliation looks and works the same in nature. After hundreds of animal studies, we realize how widespread it is. It's known of every social mammal, from rats and dolphins to wolves and elephants. They do it in different ways, though: some species groom and softly grunt, while others engage in genital frottage. Reconciliation after fights is, in fact, so universal and the benefits are so obvious that nowadays we'd be surprised to find a social mammal that *doesn't* make up after fights. We'd wonder how they keep their groups together.[27]

Male chimpanzees reconcile more readily than females. At Burgers' Zoo, male-male fights were reconciled 47 percent of the time, but female-female fights only 18 percent. These percentages correct for the rate of conflict, which is higher among males. Clashes between the sexes fell somewhere in between. Males carry their tensions on their sleeves. If one of them gets upset because his pal does something he dislikes, such as inviting a sexually attractive female, he immediately signals that something is amiss. If the other refuses to bend to his will, a confrontation may ensue. However, most of the time, they quickly reconcile. This happens even between rivals. Males are opportunistic and regularly make and break coalitions, meaning that even their greatest rival could become a future ally and vice versa. They keep all options open.[28]

Sometimes chimpanzee males disarm their rivals through lighthearted fun expressed in laughlike facial expressions and hoarse laughing pants. Like us, they use such behavior to defuse tensions. In a typical scene, three adult males are carrying out impressive charging displays. They swing from branch to branch, throw things around, and bang on resounding surfaces. In this potentially explosive situation, they test each other's nerves. But when they walk away from the scene, one of them sneaks up behind another to pull his leg while audibly laughing. This male resists and tries to free his foot from the other's grip but is now laughing, too. The third male joins in, and before long, the three big males are horsing around, punching each other in the sides, while literally letting their hair down.

Such scenes are unthinkable among females. They have far fewer open conflicts, but the ones they do have appear more intense. They're not more physical or more dangerous but perhaps more emotionally taxing. The reason for a confrontation often remains obscure. Two female chimps meet up and everything seems fine, but then they both suddenly burst out screaming at each other. As the observer, I don't have a clue as to what triggered the flare-up. Its suddenness makes me speculate that something had been brewing under the surface, perhaps for days or weeks, and that I happened to be present when the volcano erupted. This kind of eruption is rare among males, who signal hostilities and disagreements openly and easily through piloerection. Things are always "talked" out one way or another. Even if aggression does erupt, at least the air is cleared.

Another difference is that during severe quarrels between females, both parties bare their teeth and scream. This loud, piercing vocalization has many nuances, from complaint to protest, but always indicates fear and distress. It's the equivalent of human crying but sans tears. Seeing this expression in both contestants is odd compared to males, in which it signals the loser. The male who has the upper hand enlarges himself and keeps his lips firmly pressed together, whereas his opponent screams in fear and tries to move out of the way. Female fights don't show the same asymmetry. It's often hard to tell who won, and their fights rarely change anything in the rank order. Female confrontations aren't about status, and both parties scream in anguish.

Given that four out of every five female conflicts go unreconciled, it's fair to say that female chimpanzees are touched more deeply and are less willing than males to get over their disagreements. In the wild, too, females rarely make up after fights. They tend to disperse, which makes for an easy solution. It's not that females are unable to reconcile, however. When dispersal is blocked, as in one relatively crowded zoo colony, female reconciliation can be common. Female chimpanzees do reconcile, but only if they have to.[29]

For male chimpanzees, on the other hand, dispersal isn't an option, not even in the wild. Males jointly defend a territory against neighbors, which means that they need to stay united under all circumstances. They have other shared interests as well, such as their political alliances and team hunting. In general, reconciliation is tied to the importance of social relationships. This idea, known as the *valuable relationship hypothesis*, has been tested repeatedly and found to hold for both primates and other animals. Accordingly, reconciliation is most typical of parties who stand much to lose from lingering tensions. Since male chimpanzees depend more on each other than do females, it is paramount for males to mend ties.[30]

Nevertheless, all female chimpanzees are devoted to their families and have a couple of loyal friends. They need to protect these relationships and do so mostly by evading conflict. This is my peacemaking/peacekeeping hypothesis: males are good at making peace once conflict has erupted, while females are good at keeping the peace by suppressing conflict. Since males cycle with ease through fights and reunions, they don't think twice about confronting each other. Most of the time, it's no big deal. On the other hand, for females, conflict seems emotionally disturbing and nearly impossible to put behind them. The damage is so great that they develop a preemptive attitude. They take care to stay on good terms not only with those close to them but also with their rivals. They have no need to start a fight over anything minor. If a fight can't be avoided, however, females will let aggression run its ugly course.

At Burgers' Zoo, I often found Mama and Kuif grooming as if time had stood still. They were the best of friends for nearly four decades. Nothing could break their tie. I remember periods when, during power struggles among the males, Mama favored one contender while Kuif favored another. I marveled at the way they acted as if they didn't notice the other's

painful choice. Mama would make a wide detour during political com-
motions to avoid coming face to face with her friend, who had joined the
enemy. Given Mama's undisputed alpha status and her extreme temper
toward females who failed to follow her lead, her lenience toward Kuif was
an amazing exception. I never saw the slightest quarrel between those two.

Concerning women's cooperative talents, a better primate comparison than the chimpanzee might be the bonobo. Bonobo females build
a sisterhood to curb the excesses of male violence. Their ties are criti-
cally important to them, which is why they spend much time grooming.
This is also reflected in reconciliations after fights. In bonobos, neither
sex is more conciliatory than the other, and reunions are common after
female-female conflicts. Females make up quickly and smoothly, often
with intense sexual contact. One moment, two screaming females are
bashing each other, and the next they initiate genito-genital rubbing,
and it's all over. This switch may happen in the midst of a quarrel, making one wonder how deeply their animosity truly went. The grudges
that female chimpanzees hold are largely absent in bonobos.[31]

The value of relationships dictates the need for conflict management.
This is why the sexes handle conflict differently in male-bonded apes
than in matricentric ones. If conciliatory tendencies are shaped by biological evolution, think of the additional possibilities open to cultural
evolution. We are the only hominid with a balance between male and
female bonding and are the most culturally flexible to boot.

THE PEACEMAKING/PEACEKEEPING hypothesis may apply to us as
well. After all, our conflict management style resembles that of male-
bonded apes. In both humans and chimpanzees, males are the more
combative sex but also seem the quickest to smooth things over. And
in both species, females are conflict-averse. They are deeply affected by
animosities and have trouble burying them. Women have been found
to ruminate more and longer about troubled relationships than men.[32]

There are indications that many women experience greater anxiety and discomfort than do men about conflict in the workplace. This is undoubtedly why they make an effort to maintain harmony, even if only on the surface. They stay away from individuals with whom conflict is likely, they avoid situations bound to trigger interpersonal disagreement, and they downplay any issues that arise. If this proves impossible, the next best thing is to wrap criticism in diplomatic language that takes the sting out of it. Not that this is always easy. Conflict avoidance eats emotional energy. Thus, in tense environments where they can't get away from certain situations or persons, women are known to suffer more burnouts and depressions than do men.[33]

That women must be experts at peacekeeping is clear from the friendships they enjoy and their collaborative ventures, like mother-toddler groups, food preparation groups, book clubs, choirs, and the like. A growing number of businesses also have women at the helm and/or mostly female employees. Female cooperation has a long history in our species. In hunter-gatherer tribes, women set out in small groups to gather fruits and nuts on the savanna or in the forest. They also raise children together. It is hard to overestimate the importance of collective childcare in a species in which newborns have brains that are only one-third adult size. Our young are exceptionally vulnerable and needy. Hrdy characterizes us as "cooperative breeders" in that from an early age, our babies are carried, fed, and entertained by many different individuals. In this respect, we depart radically from our fellow apes, in which mothers keep their babies close for much longer. Our species has always recognized the communal social responsibility of raising children, which calls on a multigenerational network of women, girls, and men.[34]

For me, a great visual of female cooperation was the 2019 women's World Cup final, which I watched with divided loyalties as the contenders were the United States and the Netherlands. The American women won deservedly, but what struck me most was the soccer squads' team spirit. I'd love to learn how these high-powered teams operate behind the scenes, to understand women's conflict management better. In soccer, it's not just about who scores the goals but also who makes the assists. How well a team passes the ball in front of the goal shows how much they value the collective. For players to hand scoring opportunities to one another requires generosity and solidarity, which both teams demonstrated at the highest level.

In some work environments, such as hospitals, women are in the majority. The only study I ever conducted on human behavior took place in hospital operating rooms (ORs), which are pressure-cooker situations requiring intense coordination. I got involved through an anesthesiologist who had read *Chimpanzee Politics*. He said that what happens in the OR looks strikingly similar, with males vying for position, an unassailable hierarchy, and a microcosm of human social exchange, including angry outbursts. Conflict in an OR is highly problematic given the lives that are at stake. By one shocking estimate, medical errors are responsible for about one hundred thousand avoidable deaths in the United States alone. OR teams are an essential part of the equation, and signs of their dysfunction are everywhere.

Our study of human interactions in the operating room revealed that gender affects conflict and cooperation in ways that parallel primate behavior.

For instance, in one U.S. hospital, a surgeon was so displeased with an instrument that a technician had handed over that he slammed the instrument down, breaking the technician's finger. He was asked to attend an anger-management course.[35] Another hospital suspended a surgeon for unprofessional conduct, including screaming profanities at staff. Yet another hospital had to temporarily close its surgery department because a "tyrant" head had created a climate of fear that prompted staff to quit. Complaints about rude and arrogant surgeons are commonplace, and disturbing incidents occur worldwide. Hospitals naturally worry about their liabilities.

Our large university hospital's highest administrative levels granted my team permission to document what happens inside its ORs. Previous studies had been mostly of the questionnaire type, asking hospital personnel how matters went after a surgery. While this method is convenient for researchers, it is bound to produce misinformation. If you ask anyone about a conflict that has occurred, it's always the fault of someone else. An even-handed description is nearly impossible to come by. I felt that we should study people in the OR the way we study primates: by observation.

We were not allowed to film, and narrating observations would have drawn attention. So every morning of our study Laura Jones, a medical anthropologist with years of hospital experience, entered a preassigned OR. She'd sit discreetly on a little stool in a corner and take notes. She had developed a large set of behavioral codes that allowed her to key every observed interaction into a tablet. In the end, Laura recorded more than six thousand social exchanges that occurred during two hundred surgical procedures.[36]

Even though the team is there for the surgery, they sometimes spend eight hours or more together in a relatively small room, showing a wide range of behavior. Most of their social interactions have nothing to do with the medical procedure at hand. ORs play music (selected by the surgeon) while the team members gossip, flirt, joke and laugh, discuss sports and politics, exchange news, show pictures of pets, dance and sing, and get irritated or angry. Luckily, patients are oblivious to the irrepressible human sociality around them! Laura noticed conflicts occurring during about one-third of the procedures, but serious ones (involving thrown equipment or violent outbursts) in only 2 percent of them.

Most criticism is directed down the ladder from the attending surgeon, anesthesia associate, and circulating nurse to the scrub nurse. It rarely travels upward. I have heard complaints about the strict hierarchy in ORs, but can't imagine the alternative. I wouldn't want to be under the knife while a democratic team took its time to debate every critical decision. Quick action demands a stratified team. The surgeon is the alpha individual in the room. He or she is the one who will be praised if everything goes well or blamed if it goes wrong.

As for the genders, they appear equally hierarchical in ORs. We were unable, for example, to detect behavioral differences between female and male surgeons. Having read about women's and men's leadership styles, we had expected that men would be more authoritarian and women more supportive and personable. This may be how surgeons evaluate themselves or how others evaluate them, but going by observable behavior, all surgeons acted the same. Both genders were equally in charge and exhibited the same manners.

A difference did emerge, however, concerning the gender composition in the room. In terms of friendly interactions and cooperation, majority-male teams did worse than majority-female teams. This may have been due to men's boisterous behavior when they get together. Even more fascinating were gendered interactions between the alpha individual and the rest of the OR. When a male surgeon worked in a room full of women, we measured a higher level of cooperation than when a male surgeon was surrounded by men. Conversely, when a female surgeon worked in a room full of men, there was more cooperation than when she was surrounded by women. We counted twice as many conflicts if the alpha individual's gender coincided with that of most OR personnel. Since the surgeon sets the tone for OR interactions, this is what any primatologist would have predicted. The alpha position is always most meaningful within a gender. Alpha individuals feel the need to underscore their status especially vis-à-vis members of their own gender, so they may be harsher toward them. It also suggests that a female matrix similar to the male matrix steers women's attention. My anesthesiologist friend was right—the OR does resemble a monkey rock.

Productive mixed-gender teamwork requires cultural guarantees of equality. Men need to respect women in the workplace, and society must offer equal opportunities for specialized careers such as surgeon. We know how long it has taken to get us there, and how fragile these guarantees remain, but the days of a male surgeon dominating a room full of female nurses are over. Our long evolutionary history of same-gender cooperation notwithstanding, mixed-gender teams work remarkably well.

WHEN IT COMES to interaction between the genders in humans, one last sexually dimorphic trait to consider is the voice. We are a verbal species, and the voice is hugely important to us. And here I don't mean the content of what we say, but how we say it, how loudly, and with what vocal timbre.

We are so attuned to voices that they serve as individual identifiers. They do so in other animals as well. How long a voice lingers in a chimpanzee's memory struck me one day on a visit to a primate facility in Texas. My hosts informed me that they were housing Lolita, a chimpanzee I had known more than a decade earlier and had not seen since. When I went to see her, however, I was wearing a face mask. I walked up to the area where she spent her time with others, and she failed to recognize me, seeing only my eyes. There was zero reaction. But hearing my voice changed everything. After I merely said hello to her in Dutch, Lolita ran up to me with enthusiastic greeting grunts.

I don't have the low, booming voice of many men—my natural voice is rather thin and high-pitched. But I can make it sound deeper without much effort and make myself heard loud and clear. Nature has bestowed this advantage on men by lengthening their larynx. People are sensitive to voice pitch, such as when we hear a dog behind a door and can tell right away if it is a shih tzu or a Saint Bernard: the bigger the dog, the longer its larynx, hence the fuller its bark. When I say that "nature" has given men this advantage, I mean that the male voice doesn't need to sound as deep as it does. Driven by testosterone, the larynx descends at puberty in boys but not in girls. This drop, which causes the male voice to break, signals the rise in body strength. But since men's larynx is 60 percent longer than that of women, whereas their average height is only 7 percent greater, it becomes excessively long. The human male voice's timbre is much lower than you'd expect based on body size alone.[37]

Women may try to borrow the intimidating effect of a low voice, but except for a minority who naturally own such a voice, they risk sounding strained. This happened to the now-disgraced CEO of Theranos, Elizabeth Holmes, whose weird voice was debated to death on the Internet. Her voice, described in *The Washington Post* as "a deep, back-of-the-throat baritone, with a surfer inflection, a dash of seasonal allergies, a touch of robot," was preposterously low for a woman. After she was exposed as a scam artist who defrauded Silicon Valley investors, many came to believe that her low voice was as fake as her product. Perhaps she cultivated it to convey age and experience, on a par with the older male colleagues with whom she surrounded herself, such as former secretary of state Henry Kissinger, who has the most gravelly voice I know. According to co-workers, Holmes could not always maintain her adopted voice. At drinking parties, she was said to suddenly slip up and speak in a squeaky voice that sounded more natural.[38]

Only one category of people has firsthand knowledge of the typical way both genders are treated by society. Many transgender persons have lived for years as a different gender than the one they identify with. Their transition typically involves a change in clothes and hair but also body and voice. As a result, they know both sides of the coin. Their experiences, documented in informal personal accounts, confirm the worst stereotypes about the genders' position in society. It's like a trade-off between the advantages of being a man or a woman. Compared to their former lives, transgender women enjoy an increase in consideration but suffer a decline in respect. Transgender men, by contrast, enjoy more respect but meet less consideration.

After their transition, gender-conforming transwomen are treated more kindly and helpfully than they ever were as men. People smile at them in public spaces, hold a door for them, and lift their suitcases into the overhead bins on airplanes. Bystanders worry about them if they appear to be in pain or in trouble. The smile—an age-old primate appeasement signal—is directed at them more often and more easily than before. The increase in gentle treatment comes at a price, though. It reflects the view of women as vulnerable and dependent, which means that they are taken less seriously. Their voices are ignored in meetings, and they are elbowed aside in the subway. A man walking toward them expects them to cede the sidewalk. When some brave women have put this dynamic to the test by refusing to budge for oncoming men, numerous collisions were the result.[39]

Transmen report the exact opposite. Suddenly they are deprived of the friendliness, smiles, and common courtesies that they were used to as women. They are treated as autonomous beings capable of fending for themselves. With no one worried about their well-being, they get the message, *You are on your own.* One transman had a rude awakening when he left the house for the first time in his male persona: "When a woman entered a department store in front of me and just let the door swing shut behind her, I walked into it face first."[40]

On the other hand, being seen as a man brings instant authority. The transman enters a world of minimized missteps and amplified successes. Suddenly, his opinion matters. Thomas Page McBee used to have a beardless androgynous body that perplexed his professional colleagues to the point that they asked him to stay clear of important clients so as not to confuse them. All that changed following his transition:

Testosterone made my voice low. Really low. So low that I am almost impossible to hear in a loud bar or a cacophonous meeting. . . . But when I do talk, people don't just listen: they lean in. They keep their eyes focused on my mouth, or down at their hands, as if to rid themselves of any distraction beyond my powerful words.[41]

The first time he noticed everyone hanging on his lips, McBee was so astonished that he failed to finish his sentence. People patiently waited for him to continue, however. Had he been a woman, they might have jumped in, but men get a break. Not only that, men exploit their voices by addressing each other loudly, over the heads of women. They don't seem to hear women, breaking them off midsentence.

None of this is fair, of course. Worse, it's not even wise. How does it serve sound decision making if opinions are prioritized by the timbre of the voice that expresses them? It's a ridiculous standard for a smart species like ours. Let me reiterate therefore that none of the above endorses these attitudes. Instead, it highlights how deeply primate sexual dimorphism sticks in our subconscious.

Scientists have studied the impact of voice pitch by experimentally transforming it. When they played computerized male voices to young adult listeners, the voices with a low timbre were perceived as conveying higher status. Men with low voices were thought to be more likely to win a fight (physical dominance) and were also perceived as more prestigious, respected, or worth listening to (authority). In a Dutch study, young women found men with deeper voices more attractive in the same way that many women prefer physically fit men. This probably relates to men's protective role, even though the voice is a remarkably poor marker of physique. The voice is only loosely tied to body traits such as a man's bulk or the amount of hair on his chest. The male voice possibly evolved its deeper timbre as a dominance signal, for which both men and women developed a keen ear.[42]

As a university professor, I have never applied a GenderTimer (a smartphone app that measures speaking time by gender), but had I done so over the years, I would undoubtedly have found a steady increase in the amount of time women spoke at faculty meetings. One reason is the growing number of female faculty, but another is that the interaction rules have changed. If there is one group that knows, or ought to know, about implicit gender bias, it is a collection of psychology professors. Most of them disagree with the bias and try to counter its effect on discourse. Nowadays, if a woman gets interrupted by a male colleague, she's more likely to come back with something like "Hey, I wasn't finished!"

Nevertheless, studies show that in formal settings women often stay silent when men speak. After an academic seminar, for example, men ask 2.5 times more questions than women. The gender balance during Q&As improves somewhat if there are more women in the audience or if a woman asked the very first question.[43] Men trampling the discourse of women remains an everyday spectacle, though. Most recently during a televised debate between U.S. vice-presidential candidates Kamala Harris and Mike Pence, Pence kept interrupting and droning over Harris while she stayed remarkably calm, repeating over and over "I am speaking." Pence also interrupted the female moderator, who proved unable to restrain him.

The curious reality is that while our civilization values intellect, education, and experience, we still fall for crude body parameters that have no bearing on these qualities. We look down on the brute force that we believe underpins the natural order, proud to have left "might is right" behind, yet we remain stubbornly sensitive to our species's sexual dimorphism in height, muscularity, and voice. Turning this situation around will require more than a GenderTimer and a few new debate rules. A good start would be to appreciate the evolutionary roots of these biases. But while our fellow primates offer ample clues, we should also consider our species's potential for behavioral modification. We'll urgently need it if we wish to build a society where men and women cooperate on an equal basis.

11

NURTURANCE

Maternal and Paternal Care of the Young

Any scientist worth their salt loves the unexpected. That is where new insights lurk. As Isaac Asimov, the science fiction novelist, once put it, "The most exciting phrase to hear in science, the one that heralds new discoveries, is not 'Eureka!' but 'That's funny.'"

Robert Goy, director of the Wisconsin Primate Center and a pioneer of the study of hormones and behavior, once had something funny to tell me. Bob, who was a good friend and mentor, looked at me with a gleam in his eye as if he were about to divulge a little secret. "What happens if you add a baby rhesus monkey to a cage with one adult male and one adult female of its species?" he asked, then answered his own question. If both adults are familiar with infants, but neither has seen this particular one before, they'll be reluctant to touch it. After the initial uneasiness, however, it's invariably the female who responds. She will pick up the infant and place it on her belly while lip-smacking at it—a reassuring gesture. The male will barely glance at the little one. Of course, he has seen and heard it, but he acts as if it weren't there. The longer the female sits with the baby cozily clinging to her, the sleepier she will get. Holding an infant has a warm glow effect on primates.

So far, so good. But then came Bob's next question. What happens if a baby monkey is put in with a solitary adult male? The male will show the same initial discomfort and hesitation, he said, and may even retreat to a corner. Most males, however, will end up doing exactly what the female did. They'll pick up the baby and place it on their belly in the correct position, where it will soon calm down. They, too, will lip-smack while holding it tenderly, showing themselves perfect father figures.

In other words, how a male reacts to an infant depends on the presence of a female who isn't even its mother. Despite their dominance, male rhesus monkeys defer to females when it comes to the young. Bob's point was that it's not that males don't care about infants, or that they're inherently clumsy, but that childcare is female business with which they don't interfere. Moreover, they have learned to be cautious. They know that frightening or harming an infant will get them in trouble with females. Only if a male finds himself alone with a cooing and geckering infant will he take appropriate action and put the infant at ease.

In most primates, the sexes differ dramatically in the amount of care they bestow on the young. Our usual interpretation is that females are devoted to infants while males aren't. In the parlance of biology, females invest in the young's growth and health, whereas males make only a one-time genetic contribution. It often seems that way, but what if behind this black and white contrast we find tendencies that are more like shades of gray? The fact that in real life we see a sharp role division, Bob suggested, doesn't mean that males lack caretaking potential.

This is something to keep in mind while we explore the "maternal instinct," which by definition concerns females. There is much to be said for this term but also much to debate. Unfortunately, speaking of an "instinct" makes motherly care sound like the behavior of a preprogrammed robot. As if every female knows right away how to handle her newborn and will do so automatically. This is highly misleading, as

I will explain. On the other hand, there is no denying that a mother's role is tied to biology.

Mammals appeared relatively late on the evolutionary scene. They split off from reptilian and bird lineages around 200 million years ago equipped with a splendid new modus of propagation. The young grow safely inside the mother's belly and are born alive but highly vulnerable. They need warmth, protection, and liquid nutrition right away. The only viable candidate to fulfill the postnatal needs of her flesh and blood, at least early on, is the mother. Unlike the myriad animals that lay eggs and then walk or swim away before they hatch, mammalian mothers are always present when their young enter the world. Males may be around, too, but there is no guarantee. To arrange for offspring to be taken care of, evolution had no choice but to go with the female. She received feeding equipment as well as a brain that considers offspring a mere extension of herself, almost like an extra limb. In the words of the Canadian-American neurophilosopher Patricia Churchland:

> *In the evolution of the mammalian brain, the range of* myself *was extended to include* my babies. *Just as a mature rat cares for her own food, warmth, and safety, so she cares for the food, warmth, and safety of her babies. New mammalian genes built brains that felt discomfort and anxiety when babies were separated from their mother. On the other hand, the mammalian brain felt calm and good when the babies were close by, warm, and safe.*[1]

Mammalian mothers come equipped with a womb, placenta, mammary glands, nipples, hormones, and a brain designed for empathy and bonding. Their nurturing tendency doesn't always come out immediately, though, especially not in first-time mothers. It may appear piecemeal and with great ambivalence, then strengthen with olfactory cues, cries of hunger, and nursing. Most fish and reptiles don't need any of this and may even look at newborns as food, but if mammalian females don't feed their young from day one, they perish. We all descend from mothers who carried fetuses to term, produced nutritious body secretions, and were ready to lick, massage, hold, rock, and touch their young as needed for healthy growth and development.

Frequently licked rat pups grow up better socialized and more curious than do rarely licked pups, which are high-strung and nervous. Similarly, human children raised without the touch and holding of parents or their substitutes develop profound emotional disturbances. The world witnessed this sad outcome in Nicolae Ceauşescu's Romania. Its orphanages became known as the "slaughterhouses of souls" due to the disastrous consequences of contact deprivation.

Nursing of the young characterizes all mammals. It fosters an age-old emotional connection regulated by hormones and brain chemistry that is the same across species.

The way mothers bond with offspring has been likened to falling in love. But this gets the evolutionary order wrong, so we'd better turn it around. Maternal love came before the romantic variety. Female mammals of all shapes and sizes, from mice to whales, have been giving birth to helpless young for millions of years. Under the influence of a hormonal cocktail of estrogen, prolactin, and oxytocin, a pregnant female's body prepares itself for the arrival of new life. These hormones enlarge the brain's emotion hub, the amygdala, and drive care, protectiveness, and lactation. Also known as the "cuddle hormone," oxytocin is the maternal hormone par excellence. It helps induce labor, is released during breastfeeding, and fosters emotional bonding.

This whole package of physical changes is so ancient that scent remains key, even in our highly visual species. The smell of offspring has a direct pathway to a mother's brain, where it activates pleasure centers, almost like a drug. Women find their baby's smell intoxicating. They aren't bothered by poop either. In blind tests, mothers rank their own baby's feces-soiled diapers as less stinky than those of someone else's baby.[2]

All other social bonds piggyback on this ancient brain chemistry. It works for both genders, including nurturing fathers and the male-female pair-bond of some species, such as ours. When young people fall in love, they duplicate the mother-child connection. Seeing each other through rose-colored glasses, they use diminutive nicknames such as "babe," "doll," and "sweet pea," engage in high-pitched melodious "baby-talk," and feed each other as if they couldn't eat by themselves. This euphoric state goes together with elevated oxytocin levels in both lovers' blood and brain.[3]

Maternal attachment is the mother of all bonds.

IF SOCIALITY OWES so much to the love and care of mothers for their little ones, we'd better show it some respect. Evolutionary biologists, however, tend to take the way mammals reproduce for granted.

It's essential, of course, but so is breathing and locomotion. No need to make a fuss about it.

It's precisely because mammalian maternal care is ubiquitous and vital, however, that it may have served as the crucible for the evolution of social intelligence. For one thing, a mother will be better at her job if she recognizes her offspring's needs and knows what they can and cannot do. She needs to be in tune with their every little step or jump and be able to adopt their point of view. Take an orangutan mother traveling through the canopy with her dependent juvenile. Orangutans are masters at moving from tree to tree without ever descending to the ground. Due to the gaps between trees, however, travel is much easier for an adult with her long arms than for a juvenile. Young orangutans often get stuck and have to call Mom. She always returns to a whimpering youngster. She'll first swing her own tree toward the one the youngster is trapped in and then drape her body between both trees as a bridge. She'll hold on to one tree with a hand and to the other with a foot, pulling them closer until the juvenile has crossed over her body. She is emotionally engaged (mother apes often whimper when they hear their offspring whimper) and comes up with a solution that suits her child's abilities.

Taking the perspective of others has traditionally been hailed as a uniquely human ability, but it is now well documented in apes and a few other large-brained species, such as members of the corvid (crow) family. A recent study showed that apes even grasp that their perception of reality might differ from that of others.[4] They are also known for *targeted helping,* which is assistance based on an appreciation of another's predicament. Tree-bridging by orangutans is one example, but we also have experimental evidence for it. At the Primate Research Institute in Japan, the primatologist Shinya Yamamoto placed two chimpanzees side by side in separate areas. One chimpanzee had a choice between seven different tools, whereas the other needed one specific tool, to reach goodies or juice. The first ape had to look through a window at the other's situation before picking out the most suitable instrument and hand it over. That chimpanzees succeeded at this task, even though the first ape received nothing in return, demonstrated their ability to grasp the specific needs of others and their willingness to help.[5]

Mother chimpanzees demonstrate this capacity every day in the Goualougo Triangle in the Republic of the Congo. While they are fishing for termites, they respond to begging offspring by handing them a tool or letting them remove one from their hands. Not every random stick or twig has the right shape and length to extract insects. Tools picked out by the mothers are the best. Instead of letting them fend for themselves, then, mothers teach their young. They anticipate their offspring's demands by bringing extra tools to the termiting site. Teaching is yet another form of perspective taking as it requires a competent individual to appreciate another's incompetence.[6]

Let me add an anecdote that captures perspective taking in an entirely different light. At the Yerkes Field Station, I developed a special bond with Lolita, a female chimpanzee who was the star of our cognitive tests. One day Lolita had a brand-new baby, and I wanted to get a good look at it. This is hard to do since a newborn ape is really no more than a little dark blob against its mother's dark tummy. I called Lolita out of her grooming huddle in the climbing frame. As soon as she sat down in front of me, I pointed at her baby. In response, she took its right hand in her right hand and its left hand in her left hand. This sounds simple, but to do so, she had to cross her arms because the baby was clinging to her belly while facing her. Her movement resembled that of people crossing their arms when grabbing a T-shirt by its hems to take it off. Lolita slowly lifted the baby into the air while turning it around its axis, unfolding it in front of me. Suspended from its mother's hands, the baby now faced me instead of her. With this elegant motion, Lolita demonstrated that she understood that I would find the front of her newborn more interesting than its back.

All this to say that putting oneself into another's shoes, which represents an enormous leap in social intelligence, may well have started with the mother-offspring relation. This also holds for the evolution of sociality and cooperation in general. I'm convinced, for example, that the amount of ink that scientists have spent on the "puzzle of altruism" could have been greatly reduced had we considered how mothers treat offspring. Altruism is a puzzle only because of our assumption that animals have no reason to worry about others. Egoism is the way to get ahead, so why should they care about anybody else? But most animals ignore this advice. They warn others against predators, share food with the hungry, slow down for limping companions, and defend each other against attackers. Apes have even been known to jump into cold water to save a drowning companion or to chase off a formidable predator, such as a leopard, which has attacked one of them. Afterward they lick their mates' wounds and wave away the flies attracted to them. How to explain such concern for others?[7]

Maternal care, despite being both the most striking and most common form of altruism, was carefully kept out of this debate. Sacrifices on behalf of one's progeny were not puzzling enough, so their inclusion would only muddy the issue, it was felt. As a result, we went around and around about the oddity of animal kindness without ever acknowledging its ancient roots in care for the young. These roots are crucial since all mammalian rescue actions, especially in response to signs of pain and distress, follow the neural blueprint of parental care.[8]

Chimpanzees and bonobos spontaneously console upset individuals, such as those who have lost a fight. They kiss, embrace, and groom the other until he or she has calmed down. Similarly, a dog will lick and gently nuzzle a crying person or rest its head in their lap. Elephants will rumble and place a trunk in the mouth of a herd member who was startled by sudden noise. Expressions of animal empathy are increasingly recognized, and their neurobiology is shared across species. A first neuroscience study addressed the consolation behavior of prairie voles, which are small monogamous rodents. After a stressful event, one mate of a pair will groom the other. Oxytocin sprayed into the nostrils of both men and women enhances human empathy; just so, a vole's tendency to alleviate its mate's anguish has been found to depend on oxytocin in its brain. It all harks back to the first form of empathy: the body comfort that mammalian mothers offer to scared or hurt young.[9]

The only time the mother finds relief hard to provide is when she herself is the source of the discomfort. This is inevitable during weaning, which an ape mother initiates by pushing her progeny away from her nipples. For four years, the youngster could drink whenever he or she wanted, but now Mom's arms are firmly folded over her breasts. True, she will permit some brief nursing after screams of protest, but the interval between rejection and acceptance grows longer with the juvenile's age. Mother and offspring bring different weapons to this battlefield. The mother has superior physical strength, while her progeny has a well-developed larynx (a juvenile chimpanzee easily outscreams half a dozen human children) and powerful blackmailing tactics. The youngster cajoles her with pouts and whimpers and, if all else fails, a temper tantrum. At the peak of this noisy display, he or she chokes on the screams or vomits at her feet, thus delivering the ultimate threat: a waste of maternal investment. Due to their lengthy nursing period, bonobo and chimpanzee youngsters have "terrible fours" similar to our species's "terrible twos."

One wild mother chimpanzee's answer to these histrionics was to climb high up into a tree and throw down her son while, at the last instant, holding on to his ankle. The young male hung upside down for fifteen seconds, screaming his head off. Then his mother retrieved him. She did this twice in a row. There were no more tantrums that day.

I've witnessed amusing compromises, though. One youngster, five years old, took to sucking his mother's lower lip, settling for ersatz nursing. Another juvenile stuck her head under her mother's armpit, close to the nipple, to suck on a skin fold. These compromises last only a couple of weeks, however. After a while the youngster gives up and subsists on solid food, although often not without a long period of thumb-sucking.[10]

Given their nearly identical anatomies, human and ape mothers hold, carry, and nurse infants similarly. This is why zoos sometimes invite human mothers to demonstrate to naïve apes how to breast-feed, and why zoo caretakers and regular visitors often tell me stories of apes being extremely curious about human pregnancies and newborns. They follow the process closely. One woman related that after she gave birth, she went to see the zoo gorillas, wheeling her carriage to the edge of the moat. She was greeted by a gorilla whom she knew well and who was carrying her own newborn. At first they stood gazing at each other, but then the gorilla patted her belly while staring at hers, which she patted in return. "We were mothers, joined," the woman said.

One final parallel is the tendency to cradle infants in the left arm. This unconscious preference marks about four out of five human mothers. The left-side preference is specific to infants and dolls and doesn't apply to other objects being held. Since the same side bias is known in ape mothers, it is unlikely to be cultural. There are several theories about it, such as that it brings an infant closer to the mother's heart, where it can hear its beat. Or that it keeps the dominant right arm free for other tasks. However, the best-supported idea is that objects in the left visual field are perceived mostly by the right side of the brain, due to the crossover of visual information in the optic chiasm. Given that the right brain hemisphere processes facial emotions, cradling a baby on the left promotes the emotional connection.[11]

Things aren't entirely up to the parent, though. Babies aren't passive, and most of them prefer to nurse on the left. This left-nipple bias, too, marks both humans and apes.[12]

THE FIRST TIME I realized how finely attuned apes are to their offspring was when a deaf chimpanzee named Krom kept losing hers. Mother apes listen for small, barely audible sounds of contentment and discomfort from their babies to know how they're doing. But deaf Krom couldn't hear these sounds or any louder calls. If she sat down on her baby, she didn't even react to its screams of protest. The feedback chain was broken. Despite being a great mom with strong nurturing tendencies, she failed. We removed her last offspring before there would be another sad ending.

Her infant, named Roosje, was given for adoption to Kuif, a chimpanzee obsessed with infants but with insufficient lactation. We had noticed that if Krom neglected her infant's cries, Kuif sometimes started crying, too. We were able to teach Kuif to bottle-feed Roosje, which shows that chimpanzee maternal behavior is flexible enough to add a completely novel technique. Kuif even learned by herself to remove the bottle if Roosje needed to burp, something we'd never taught her.

The fact that maternal behavior can be learned is why the term *instinct* falls short. Even natural nursing isn't as self-evident as it seems. For example, human neonates have trouble with unassisted nursing even though they are born with the same rooting and sucking reflexes as all mammals. Guided by the smell of the breast, the baby tries to latch on. The nipple touching the roof of its mouth triggers rhythmic sucking cycles. But if the nipple stays out of reach, this can't happen. Human breasts are relatively large and swollen, making the nipple like a mini Mount Everest. In other species, the young only need to walk up to a mother who is lying on her side. Or they have an udder hanging right above their head. Human mothers have to position their babies correctly—otherwise nursing won't take place. In addition, the areola needs to be compressed for milk to be expressed, so babies need to close their mouth around more than the tip of the breast. The role of reflexes notwithstanding, successful nursing takes a lot of learning on the part of both mother and child.[13]

There are other complexities to mothering, such as how to carry a baby, when and how to react to its cries, how to clean it, how to reassure it if it is upset, and later on how to educate it. Nature dictates none of this. These skills are acquired early in life by observing and imitating competent mothers and helping them care for their babies. Maternal traditions are passed on from generation to generation. There is no way this would happen if females weren't hugely attracted to newborns. In the same way, it is unlikely that we could have taught Kuif how to handle a bottle if she had been indifferent to Roosje. Motivation is key.

Quite a few studies have measured children's preference for pictures of adults versus infants or observed their reactions to actual babies, such as one that experimenters left alone in a waiting room. From preschool age onward, girls are more interested in babies than boys. They talk to them, kiss them, and try to hold them. When asked to take care of a baby, girls do so more eagerly than boys. To see if the mother encourages this difference, one study observed five-year-olds interacting with a new member of the family. Girls attended to and nurtured their baby sibling more than did boys. The mother, who was present, did not instruct her daughter to do so, however. She talked equally with sons and daughters about baby matters.[14]

Attraction to babies is echoed in attraction to the oldest toy known to archaeologists: the doll. While boys can turn almost any random object into a sword or gun, often over parental objections, girls who lack commercial dolls get creative with materials at home. They follow an ancient tradition of self-made dolls, such as the cornhusk dolls of Native Americans and Inuit dolls made out of soapstone and animal fur. The girls' imagination fills in the blanks, as the Russian developmental psychologist Lev Vygotsky noted: "a bunch of rags or a piece of wood become a small baby in a play because they allow performing the same gestures as when carrying a baby or feeding it."[15]

A similar kind of imaginative play has been observed in our closest relatives. As discussed in Chapter 1, apes often turn inanimate objects into dolls, which they treat like infants. While chimpanzee Amber schlepped a soft broom around, there are reports of her wild counterparts in Uganda carrying wooden logs in the forest. Whereas males look at these logs as playthings, females take a nurturant attitude. They make the log ride on their back, hold it tight while asleep, or build a cozy nest for it.[16]

Young females stand much to gain from early experience with babies or pretend-babies. While young male primates have their bois-

terous rough-and-tumble games to prepare for a life of status competition, young females are busy acquiring maternal skills. I realize how
stereotypical this sounds, but also feel that this term is thrown around
a bit too casually. Merriam-Webster's dictionary defines *stereotypical* as
"conforming to a fixed or general pattern or type, especially when of
an oversimplified or prejudiced nature." Characterizing children's games
this way implies that all they do is follow some societal ideal. However,
the biological reality is that the sexes reproduce differently and that the
young prepare for this future. It works the same for all animals. This is
why young billy goats playfully butt heads all day, why female dogs drag
fluffy toys around the house like puppies, why young male weaverbirds
construct play nests, and why young rats play-mount each other. All of
this is done in fun, but these behaviors will one day decide who gets to
spread their genes. Children's games follow the same playbook.

If the interest of girls in infants and dolls were purely cultural, it
should vary from place to place and from era to era. But it hardly does.
It has been known since at least the ancient Greeks and Romans. Observations in ten different cultures found girls to be more nurturant and
involved in household tasks while boys played more often away from
home. Most of these studies were conducted in the 1950s, before Western television and movies overtook the world, in countries as diverse as
Kenya, Mexico, the Philippines, and India. The American psychologist
Carolyn Edwards concluded, "Clearly, girls perform more infant care
and are more involved with infants than boys in many subsistence-based
societies in which busy mothers recruit help from older children." Even
in cultures in which men are intimately involved in domestic life, girls
handle infants more than do boys.[17]

Edwards offers self-socialization as an explanation. Socialization isn't
always imposed by society—it may come from the child itself. Since
both boys and girls prefer same-sex company, girls spend more time
around women. Combined with their fascination with infants, this
automatically gets them involved with childcare. But there is more to it,
because Edwards notes the obvious pleasure girls take in infant-related

tasks. They volunteer for them. Their interest in these tasks is one of the most consistent cross-cultural gender differences.

Young female primates are as infant-obsessed as girls, whereas male interest in infants reflects an almost technical curiosity rather than a nurturing tendency. Young male chimpanzees often carry babies in an awkward manner without letting them cling to their bodies as ape babies love to do. I have watched in horror as young males inspect a small infant by stretching its limbs to the limit, sticking their big fingers down its throat, or making it the object of a tussle with a male peer. Undeterred by the baby's vocal protest, they evade the mother's frantic attempts to get her child back. Understandably, most primate mothers are reluctant to let a young male walk off with their baby unless he has proven to be caring and careful. Such males do exist, but they are generally a bit older and more experienced. With young females, the mother at least has a guarantee that the baby will be treated gently, watched over, and returned in time for nursing.

We have known this sex difference for over half a century. Young female primates like to get their hands on "something that squiggles," as one fieldworker put it. A 1971 report by the American primatologist Jane Lancaster described wild vervet monkeys in Zambia: "By the time an infant is six or seven weeks old, it is actually spending a good proportion of its waking hours in the company of juvenile females. Mother vervets often take advantage of this and go off on their own to feed." Lancaster contrasted this response with the response of males: "No male of any age was seen to direct any maternal behavior such as hugging, carrying, or grooming toward a newborn infant."[18]

In most monkeys studied, immature females interact with infants three to five times more often than do immature males. The females' behavior is known as *allomothering*—mother-like care of dependent young by an individual other than the mother. It helps them develop maternal skills. The primatologist Lynn Fairbanks investigated this in another study on vervet monkeys. She observed a large cohort of first-time monkey mothers whose history she knew since birth. Fairbanks

wanted to determine the survival rate of their offspring. Did it help them
if they had spent hours taking care of other females' offspring while
younger? It did indeed. Mothers who had done so had a lower infant
mortality than those without this early experience.[19]

Monkeys who grew up separate from mothers with infants neglect
their firstborns. They have no clue what to do and won't even pick them
up. This is also common among apes at zoos who lack a motherhood
tradition. It is crucial to introduce a female with solid maternal experience to demonstrate to them how it's done.[20] It works like this for many
mammals, including the ubiquitous "aunties" who act as babysitters for
the calves of elephants, dolphins, and whales. And even though we tend
to think that rodent maternal behavior must be inborn, for them, too,
mothering begets mothering.

Putting a camera in the dens of mice has taught us that experienced
mothers try to keep juvenile virgins nearby. If one of them leaves the den,
the mother will chase her down and shepherd her back. She teaches the
virgin how to carry pups back to the nest by dropping a pup and retriev-
ing it right in front of her. Or she deposits a pup before a young female
as if tempting her to pick it up. Young mice with this kind of exposure
learn how to handle pups more quickly than do those without.[21]

It's time, therefore, to stop calling the passion of girls for infants and
dolls stereotypical. Human behavior that is found all over the world and
shared with many other mammals isn't explained by prejudice and gender expectations, even though both may contribute. It goes deeper than
that. Biology is involved, and for a good reason. Since maternal skills are
too complex to be left to instinct, evolution made sure that the gender
that needs them most is eager to get maternity training.

A tendency that is functionally tied to an age-old mode of reproduction isn't stereotypical but *archetypical*.

Now LET'S RETURN to the observation that male primates, even in
species with little or no paternal care, are hardly indifferent to infants.

Under some circumstances, males will hold and nurture them and demonstrate an impressive caretaking potential. Moreover, this potential isn't limited to primates. For example, male rats aren't known to care for pups, yet they will do so if left alone long enough with them. The same is true for chickens, as Charles Darwin already noted in his diary. He recognized that a capon (a neutered rooster) "will sit upon eggs as well as & often better than a female." Darwin speculated that a "latent instinct" for nurturance lurked inside the male brain.[22]

This instinct is on full display in the many bird fathers who tenderly care for and fiercely defend their young but also in a few primates, such as marmosets and tamarins. Males of these little South American monkeys are heavily involved in the transport and care of the twins that females give birth to. The American primatologist Charles Snowdon studied cotton-top tamarins for his entire career. I regularly visited his colony not

Cotton-top tamarins are squirrel-sized South American monkeys with highly developed paternal care. The twins born to these monkeys are carried around more by the father than by the mother.

far from the Wisconsin Primate Center. Snowdon found tamarin fathers to be highly competent and caring. Their investment is so costly that they lose weight when they carry the young on their back. The mother's main investment is gestation and lactation. She carries the young only during nursing, leaving them on their dad the rest of the time. He generously shares solid food with the twins to get them ready for independent foraging. Already during his mate's pregnancy, the father goes through hormonal changes. He has increased levels of typically female hormones that stimulate bonding, such as estrogen and oxytocin. He also fattens up in compensation for the weight loss that awaits him.[23]

These monkeys are quite distant from us, however, which makes them less relevant to human evolution. Closer to us are the gibbons and siamangs of Southeast Asia. They are best known for beautiful coordinated singing by mated pairs high up in the treetops. The singing serves bonding and keeps neighbors out of their territory. Males and females divide the task of offspring care, with the male often carrying their singleton as well as playing and sharing food with it.[24]

Turning to our nearest relatives, the great apes, it may seem at first that males are uninvolved in offspring care, but this is not entirely the case. True, they rarely carry the young or help them find food, but they do protect them. When gorillas or chimpanzees cross the road in Africa, for example, a large male may position himself like a policeman in the middle of the road to stop village traffic. He will stand there patiently waiting until all the members of his group have sauntered across. [25] Since male gorillas are hyperprotective, Western hunters in the old days usually brought back the skins, heads, and hands of adult males. Males bluff-charged the hunters so as to give their family time to escape, and they were shot. Nowadays, thankfully, the same defensive actions result only in lots of pictures of imposing chest-beating males.

The most extraordinary act of protection by male chimpanzees I ever witnessed was during the reintroduction of Kuif and Roosje to the Burgers' Zoo colony. Having kept those two apart for weeks for bottle training, we had noticed the hostility of our young alpha male, Nikkie.

Once when Kuif walked past Nikkie's night cage, he grabbed through the bars at Roosje, who was clinging to her. Kuif jumped away with a sharp yelp. This brief interaction worried us. The last thing we needed was one of those gruesome infanticide scenes such as reported from the field. Roosje might be torn to shreds. Since I had held her for weeks, helping Kuif feed her and feeding her myself, I was far from the dispassionate observer that I like to be.

Since only Nikkie reacted this way, we decided to conduct the introduction in stages and to release Nikkie last. Outdoors, most colony members greeted Kuif with an embrace, stealing glances at the baby. Everyone seemed to be keeping a nervous eye on the door behind which Nikkie waited for his release. Chimpanzees know much better than we human observers do what to expect from each other. In the melee, we noticed that the oldest two males never left Kuif's side.

When we let Nikkie onto the island, about an hour later, these two males positioned themselves about halfway between Kuif and the approaching Nikkie, with their arms draped around each other's shoulders. This was a sight to behold, given that these two had been arch enemies for years. Here they were, standing united against the young leader, who approached in a most intimidating manner with all his hair on end. Nikkie broke down when he saw that the other two were not going to budge. Kuif's defense team must have looked incredibly determined, staring down the boss, because Nikkie fled. Much later, he approached Kuif under the watchful eyes of the other two males. He was nothing but gentle. His intentions will forever remain shrouded in mystery, but the animal caretaker and I hugged while sighing with relief.

Male chimpanzees sometimes do more than protect the young. Their nurturing capacity manifests itself during emergencies in the field. After Tia, a wild female in Fongoli, Senegal, lost her infant to poachers, researchers managed to confiscate the baby ape and return it to the group. Mike, an unrelated adolescent male, who was too young to be the infant's father, picked it up from where the scientists had left it. He knew to whom it belonged because he carried the baby straight to Tia. Mike must have noticed how much trouble Tia had moving around after she'd been mauled by the poachers' dogs because for two days he carried the baby during group travel with Tia limping behind.[26]

Even more remarkable are full-blown adoptions of unrelated young, which is the biggest investment of all. Boesch lists at least ten adoptions by wild male chimpanzees that he observed over three decades in Taï Forest. They occurred after the sudden death or disappearance of a juvenile's mother. In 2012, Disneynature released a popular movie *Chimpanzee* that captured how Fredy, the community's alpha male, took Oscar under his wing. This documentary was based on real events. When Oscar's mother died of natural causes, the camera crew stayed around even though the prospects for little Oscar looked bleak. Fredy followed the pattern of other adoptive males, who allowed the youngster to sleep in their night nests, protected them against danger, and diligently searched for them when lost. While cracking nuts with stones, they'd share the kernels. Some cared for the youngster for at least a year, and one male did so for over five years. According to DNA samples, adoptive males were not necessarily related to their charges. Oscar was lucky.[27]

At yet another field site in Kibale National Park, in Uganda, scientists witnessed an outbreak of respiratory disease that killed no fewer than twenty-five chimpanzees. Multiple orphans were the result. Since chimpanzees remain dependent on their mother for at least ten years, the usual outcome for young orphans is death. Four of them, however, were postweaning and lucky enough to have an adolescent brother. The primatologist Rachna Reddy followed these sibling pairs for over a year and found the older brothers to be extremely vigilant and responsible. The siblings traveled together, frequently groomed each other, and provided reassurance when one of them was frightened. Brothers defended their sibling against aggression and sometimes cried while searching for them if they got lost. Like a mother, they would never move on without first checking on their sibling to make sure that they were following. This attentiveness is most remarkable given the tough social life of male adolescents, who face an uphill battle trying to enter the male hierarchy.

Younger siblings often sought comforting body contact. Reddy describes how Holland, a seven-year-old male, did so with his seventeen-year-old brother, Buckner: "Holland would regularly sit so that his shoulder touched Buckner's and often, while Buckner sat upright, Holland pressed his own back into Buckner's chest or shoulder, occasionally whimpering. This continued for at least 8 months after their mother's death."[28]

Clearly, male chimpanzees possess a well-developed fatherly potential even though it's seldom expressed. We know less about bonobo males, but I have so often seen them play tenderly with infants and juveniles that I have no doubt that they, too, have this potential. One adoption was witnessed by the Japanese primatologist Gen'ichi Idani in the Democratic Republic of the Congo. Idani hand-reared a rescued baby bonobo, named Kema, who had lost her mother to poachers. Every day for two months, Idani took Kema into the forest to introduce her to a wild group. One day he left her behind. Revisiting the area the next morning, he discovered Kema in the nest of an adolescent male. The male held her while she clung to his belly. Kema successfully integrated into the wild group.[29]

Male bonobos can be highly protective. A striking example occurred at the San Diego Zoo at a time when its enclosure still had a wet moat. The keepers had drained the moat for scrubbing, then went to the kitchen to turn on the water valve to fill it up. Before they could do so, however, they were rudely interrupted by Kakowet, the alpha male. He appeared in front of their kitchen window, screaming and waving his arms. As it turned out, several young bonobos had jumped into the dry moat to play but were unable to get out. If the water flow had not been halted, they would have drowned.

Kakowet's anxious intervention demonstrated his ability to take another's perspective and recognize their circumstances. But more practically, it also showed that he knew who controlled the water supply.

After his alarm, the keepers descended into the moat with a ladder. They got all the bonobos out except for the smallest one, who was pulled up by Kakowet himself.

HUMAN MALES ARE DIFFERENT. Going beyond basic protectiveness and a caretaking potential, they evolved to provide actual family support. Our males are much more fatherly than those of many other primates. We don't know how and when this began, but it may have been when our ancestors left the forest and entered drier, more open terrain.

Don't believe those killer-ape stories of Robert Ardrey and others about how our ancestors ruled the savanna as top predators. Our fore-

Humans are unique among hominids by having families with direct male involvement in childcare.

bears were *prey*. They must have lived in constant fear of pack-hunting hyenas, ten different kinds of big cat, and other dangerous animals. Both the lions and the hyenas were larger then than they are today, while our ancestors were smaller than us. The transition out of the relative safety of the forest must have been prolonged, gradual, and extremely stressful. Ardipithecus, which lived 4.4 million years ago, still had feet more suitable for climbing than walking. This ancestor probably didn't like staying on the ground at night. Equipped with a prominent grasping great toe, she slept in the safety of the trees like our ape relatives.

In this scary place, females with young were vulnerable. Unable to outrun predators, they could never have ventured far from the forest without male protection. Perhaps bands of agile males defended the group and helped carry juveniles to safety during emergencies. This might never have worked, though, if they'd kept the chimpanzee and bonobo social system. Promiscuous males cannot be relied on for parental commitment. To get males more involved and stay nearby, society had to change.

Human social organization is characterized by a unique combination of (1) male bonding, (2) female bonding, and (3) nuclear families. We share the first with chimpanzees, the second with bonobos, and the third is ours. It's no accident that people everywhere fall in love, are sexually jealous, seek privacy, look for father-figures in addition to mother-figures, and value stable partnerships. The intimate male-female relationships implied in all this are part of our evolutionary heritage. I believe it is this pair-bond that sets us apart from the apes more than anything else.

Initially, males may have acted mostly as protectors and transporters of the young, but at some point, they began to share food with females with whom they had mated in the past. It could be that males demanded exclusive faithfulness from these females in return, but I suspect a more fluid arrangement. Today we are keenly aware of paternity and genetic kinship, but this is recent knowledge. Our ancestors most likely didn't think this way, and males may have tied provisioning and care only vaguely to their sexual history. Even today, the majority of Amazonian cultures consider children to be the product of multiple encounters between a mother and all the men she slept with.[30]

Whatever the awareness of fatherhood and the precise sexual arrangements, drawing males into family life had enormous benefits. Instead of care for the young boiling down to the mother's abilities, males brought home highly prized meat and began to help out with offspring care. This made it possible to reduce the interval between births from the five to six years of our ape relatives to the three to four years of modern hunter-gatherers. Humanity began to speed up reproduction so that some families might count ten or more children, which is physically impossible for the apes. As an ape mother travels through the trees while carrying her youngest and keeping an eye on older offspring, her family size is seriously constrained. Given the planet's current overpopulation, humanity's breeding success has been a mixed blessing, but at its root we find increased paternal engagement.

It is unlikely that male ancestors provisioned all females and children equally. They must have felt an obligation to certain women and their children. It may have been more than one, but the number was small enough that some children became special to them. Having the same potential for paternal care as all primates do, males became emotionally attached and committed to these children. The amount of care they provided varied with their precise ecological circumstances, but the tendency and capacity to do so became well entrenched in our lineage.

This doesn't mean that males care for the young in the same way as females. For one thing, a difference in empathy comes into play. While this is not the place to go over the vast literature on human empathy, one recent review put it in a nutshell as follows: "many studies converge on the conclusion that there is a female superiority in empathizing." I must add, though, that this conclusion applies mostly to the emotional side of empathy. Empathy is typically divided into two layers. Emotional empathy relies on reading body language, such as facial expressions, and on being affected by another's emotional state. This is the oldest, most basic layer of empathy, one that we share with all mammals. The second layer, which is more cognitive, develops on top of it. It takes the other's perspective by imagining their situation. Generally, women have an advantage in emotional empathy, but their cognitive empathy is similar, perhaps identical to that of men.[31]

Since the two layers are often conflated, human empathy research doesn't always find an unambiguous gender difference. If it does, however, it's always women who have more of it, never men. Another problem is the reliance of modern psychology on questionnaires and self-report. It must be clear by now that I prefer measures of actual behavior. One of the very first to collect those was the American psychologist Carolyn Zahn-Waxler, whose team visited homes and asked family members to feign sadness (sobbing), pain (crying "ouch"), or distress (coughing and choking), to find out how young children responded to them. They discovered that children between one and two years of age already comfort others. This milestone in their development comes well before language: an aversive experience in someone else draws out empathic concern, such as patting, kissing, and rubbing the victim's injury. These reactions were more typical of little girls than of little boys.[32]

It is hard to find comparable data on human adults, but one recent study worked with surveillance camera footage of the immediate aftermath of store robberies in the Netherlands. This is the moment when the police walk in to inspect the premises and write a report. Some of the victims, especially the business employees, had been subjected to physical force or threatened with a weapon. All of them were upset. The video analysis focused on comforting body contact in the store, such as touching and hugging. Female bystanders were almost three times as likely as males to console robbery victims. One explanation is that it is more acceptable for women to make physical contact, but another is that women show more concern about the well-being of others.[33]

The difference in empathy and nurturance between men and women is supported by neuroimaging research. When subjects watch emotionally charged images and answer questions about another's situation, women seem to erase the emotional boundary between themselves and the other, whereas men apply their intellect to grasp the other's situation. Women's brains show increased activity in emotion-related areas, such as the amygdala, whereas men put their prefrontal cortex to work.[34]

Parental care shows a similar gender difference in the brain, but with a twist that should interest anyone who'd like to see more equality in this domain. For my postwar European generation, fathers were emotionally distant figures, barely involved in daily childcare. They might hold our hand when we crossed the street or admonish us if we had done something wrong, but that's about it. With the ongoing expansion of men's domestic role, science is interested in knowing how it impacts their brain. The human brain is enormously flexible, a phenomenon known as *neuroplasticity*. The connection between brain and behavior is a two-way street. Not only does the brain make us behave in certain ways, it rewires itself as a result of our circumstances and behavior. For example, taxi drivers have an enlarged hippocampus due to their reliance on spatial memory, and people who learn a second language or master an instrument grow more gray matter. The brain is modified by the demands we place on it.

Parental care is a good example. The Israeli neuropsychologist Ruth Feldman has shown that the brain responds in gender-typical ways when parents watch their children. Mothers rely more on the emotion centers and fathers more on cognitive areas related to problem solving. These differences are by no means set in stone, though. Depending on how much responsibility they shoulder for childcare, men's brains change. In some couples, the wife acts as the breadwinner and the husband as the primary homemaker. There are also gay couples who have adopted children and motherless families headed by a father. In these families, fathers are much closer to their children and more involved than most men. They worry about their kids every day and have to be on standby when they are sick or in trouble. Feldman found increased oxytocin levels in the blood of these fathers and a more active and better-connected amygdala in their brain. Neurologically, their brains had taken on maternal characteristics.[35]

The parental style of the majority of fathers, however, remains quite different from that of mothers. Fathers engage more in wild play and roughhousing or take their children on daring outdoor adventures. Masculinity doesn't stand in the way of men being good caretakers. On the contrary, the more men fit the stereotypical definition of "manliness" (adventurous, dominance-oriented, competitive), the higher the rating that observers assign to their parenting attitude to a baby daughter or son.[36]

The anthropologist James Rilling, who studies human fatherhood, believes that fathers serve a special function in the development of their children:

Fathers tend to specialize in preparing children for life outside the family. Fathers are more engaged in unpredictable behavior that destabilizes a child, and the child has to learn how to respond to that. It may help to develop resilience, an important trait since not everybody is going to treat you as well as your mother does.[37]

Rilling found that fathers, after the arrival of their first child, experience not only a rise in oxytocin but also a decline in testosterone levels. They are shifting away from the risk seeking and mate hunting of younger men toward a deeper commitment to their families. These hormonal changes should shatter any myth that men can never be good caretakers because they are not biologically "primed" for it. Sitcoms and comedy shows reinforce this view by mocking dads as clumsy and clueless. In contrast, research such as that of Feldman and Rilling depicts human fathers as perfectly capable of emotional involvement in childcare. It is part and parcel of our species's biology.[38]

THE WAY FATHERHOOD impacts the brains of men parallels the evolutionary changes that turned male cotton-top tamarins into perfect
fathers. The main difference is that all tamarins raise their offspring this
way, whereas the human father role is optional. A father's contribution
varies from culture to culture in sharp contrast to a mother's contribution, which is a human constant thanks to its ties to biology.

Despite these ties, a woman can have an utterly fulfilling life without
motherhood. I speak from experience since my wife and I are childless by
choice. I don't look at having children as a woman's obligation or destiny.
Nevertheless, there are people, including most male thinkers of the past,
for whom the chief reason women are on earth is to make babies. As it is
sometimes put, men are here to produce and women to reproduce. Even
the anthropologist Margaret Mead expressed hostility to women who
"disavowed" childbearing. This was well before we had the Pill, however,
when the role division between the genders was still largely inescapable.
Only when average family size began to drop did this division lose its
grip on society and could women begin to view motherhood as a choice.[39]

Mothers have never been the sole caretakers, though. Apart from col-
laboration with fathers, our species has other "helpers at the nest." This
is a biological term for nonparental supporters of the young, such as adolescent birds that stick around to help their parents feed the next batch of
chicks. In her book *Mothers and Others,* Hrdy presents humans as "cooperative breeders" with lots of helpers, or alloparents:

> *The recognition that a child's survival depended not just on staying in contact
> with his mother or provisioning by his father but also on the availability,
> competence, and intentions of other caregivers in addition to parents is usher-
> ing in a new way of thinking about family life among our ancestors. Without
> alloparents, there never would have been a human species.*[40]

The first hints of this kind of cooperation can be seen in other
primates. For example, chimpanzee and bonobos sometimes act as

"midwives" to a pregnant female. I witnessed it once when a female exceptionally gave birth in the middle of the day. Most births occur at night when no one is watching, but one day chimpanzee May gave birth in the group. She stood half upright with her legs apart, while lowering one open hand between them in order to catch the baby when it popped out. Next to her stood her best friend, Atlanta, who adopted the exact same posture. Atlanta was not pregnant but mimicked May. She, too, stretched her hand between her legs, her *own* legs, where the gesture served no purpose. Or perhaps it was the other way around and Atlanta was instructing May, explaining *This is what you should do!* Other females closely followed the birthing process and kept May's behind clean. Similar instances of assisted parturition have been observed in bonobos.[41]

Moreover, in monkeys with extensive female kinship networks, such as macaques and baboons, grandmothers make a huge difference. They fiercely protect their daughter's offspring, play with them, groom them more than anyone else in the troop, and give their mom a break. Offspring who have a supportive grandmother around are more likely to venture away from their mother and reach independence earlier.[42]

In human society, too, the most crucial alloparent is the grandmother, especially the one on the maternal side. According to the *grandmother hypothesis,* this is why evolution gave us menopause. We are the only primates in which the female life span stretches well beyond her fertile years. Normally, this wouldn't make much sense. Why not go on making babies until the last drop? A chimpanzee female still walks around with offspring on her back at such an age that we, human observers, take pity on her. She is getting too frail for the load, the nursing demands, and the tantrums that go with them. In our species, older women are never in this situation. Hormonal changes curtail their reproduction at a time when they still have decades to go. This evolutionary "innovation" makes us the only primates in which about one-third of adult females are beyond childbearing age.

Recently, we have learned that some matriarchal, long-lived whales, such as orcas and belugas, also experience menopause. Grandmother whales increase the survival of their grand-offspring by passing on freshly caught salmon to young calves and guarding them on the ocean surface while their mother takes off for a deep dive.[43]

The grandmother hypothesis explains menopause as a reproductive strategy. Its developer, the anthropologist Kristen Hawkes, believes that the best way for older women to advance their genetic legacy is to help their daughters raise children. It's superior to attempts to raise offspring on their own. In her work with the Hadza people of Tanzania, Hawkes had noticed how incredibly productive "old ladies" were in gathering food for their families. She developed her ideas about their supportive role from there. Anthropological studies support the grandmother hypothesis, as do historical records of preindustrial societies, such as in Finland and Quebec. These records show that daughters who have their mothers around are more successful at raising children.[44]

Other primates may have a less extensive support network, but the wider community is far from indifferent to a mother's situation. Motherhood is recognized and respected by everyone. When a young female monkey becomes a mother for the first time, she gains in status. As a juvenile or adolescent, no one took her seriously, and she was commonly chased away from food or water. Carrying a newborn, however, brings instant respect and tolerance. She's suddenly allowed to eat or drink right next to the higher-ups, at least for a while. It's also remarkable how eager others are to sit with and groom a new mother as if she were the hottest item around. I know bonobo groups in which new mothers are recognizable by naked patches of skin due to all the overgrooming.[45]

Recognition of motherhood is also visible in the reaction to an infant's death, such as when at Burgers' Zoo a female chimpanzee had a stillbirth. The day that happened, the whole colony, including individuals who weren't even close to her, sympathized by frequently kissing and embracing the bereaved mother. The change lasted longer, however. For at least a month, the colony showered her with greater affection than usual.[46]

Like humans, other primates surround motherhood by expectations, such as that a female will feed and defend her young. As soon as primates hear a youngster scream in distress, all heads turn to the mother. She is the one to swing into action. This expectation doesn't exist for males, which is why male chimpanzees in Taï Forest were upset when one among them adopted a parental role. Boesch describes how Brutus ran into resistance after adopting an orphan named Ali. Since Brutus was the best hunter of monkeys in the community, he was often in possession of meat:

> *Brutus generously shared meat with many different females and some males, but never with subadult individuals as usually it is the mother who shares with them. Since the adoption of Ali, however, Brutus also shared with him and this was the cause of constant squabbles, as the adult meat beggars did not approve of the favored treatment shown to the youngster. But Brutus kept on sharing with Ali, and even handed him some of the most highly prized pieces.*[47]

If social expectations are so important, we must apply the gender concept to apes as well. They are no strangers to social norms. Some patterns of behavior are accepted, while others violate the rules and arouse protest. By acting like a good father figure in a society in which this role barely exists, Brutus was going against the grain, and the others let him know it. They objected to his violation of typical male behavior. Similarly, Robert Goy's earlier monkey example highlights what is expected from a male facing a new infant. Even though he is perfectly capable of picking it up, he regards this as a female job.

Social arrangements are sometimes more rigid than the biology behind them. While it is always unwise to ignore biology, it is a simplification to attribute existing social roles to it. Modern knowledge of animal and human behavior indicates a more flexible arsenal of responses than is often assumed.

12

SAME-SEX SEX

Animals Carrying the Rainbow Flag

At the Kyoto Aquarium in Japan, the romantic partnerships among its penguins are so complicated, with so many breakups and new liaisons, that an elaborate flowchart is needed to keep track.

The chart features each penguin's portrait and name, with two-way arrows indicating romance between two individuals and one-way arrows for unrequited love. Red hearts designate happy couples, while blue broken hearts mark terminated affairs. Breakups are common, often resulting in a loss of appetite in both parties involved. There are also love triangles and cases in which a bird reserves his or her flirtatious behavior, such as dramatic head-shaking, for a particular human staff member. The flowchart is one of the most popular items on the aquarium's website as it allows everyone to follow the latest developments on the penguin mating market.[1]

The majority of partnerships are heterosexual, but some are homosexual. Due to its human baggage, *homosexual* may seem an odd clinical term to apply to animals, but the contrast between the Greek prefixes *homo* (same) and *hetero* (different) is convenient enough that they are often used. At the aquarium, it started with one romantic BL (which in Japan stands for "Boys' Love") between an older and younger male until both of them fell hard for the same female. Penguin love life is about as complicated as its human counterpart.

There was, of course, a time when we weren't allowed to reference animal homosexual conduct. It was too shocking to think about. That it occurs in penguins has been known for over a century, though. The very first report described their behavior as "depraved" and was available only privately, to keep it hidden from a wider audience.[2]

All this changed in 2004 when *The New York Times* drew attention to two male chinstrap penguins at the Central Park Zoo in New York, who hatched an egg together. Named Roy and Silo, the birds first tried to hatch a rock as if it were an egg. This gave the caretakers an idea, and

Two male chinstrap penguins at the Central Park Zoo, in New York, helped bring public attention to animal homosexual behavior and bonding. Roy and Silo raised a chick from a fertile egg that caretakers had placed in their nest.

they offered them a fertile egg from another couple. The female chick raised by Roy and Silo was named Tango and became the inspiration for a children's book, *And Tango Makes Three.* The book was banned from public libraries across the United States for not being age-appropriate, but it turned into a bestseller nonetheless. In the years that followed, penguin sexual orientation became a topic of political debate and even public protest.

This came to a head in 2005, when the Bremerhaven Zoo in Germany sought to breed endangered Humboldt penguins. They decided to separate their male pairs and repair them with females brought in for this purpose. The zoo declared that the male-male ties were "too strong" for its breeding program, since they kept the males away from the females. Some gay organizations disputed the move as an attempt to change the birds' sexual orientation by means of "the organized and forced harassment through female seductresses."[3]

The enthusiasm of the gay community for the penguinian equivalent of homosexuality is understandable. But it is also a bit surprising given prevailing gender theory, which often touts us as capable of transcending biology. This is why we have genders, whereas animals have mere sexes. But while we often keep biology at a distance in relation to gender, we positively embrace it when it comes to sexual orientation and transgender identity. Here we eagerly explore genetic differences and the role of hormones and the brain. The same American Psychological Association that calls gender a social construct defines sexual orientation as "one's enduring attraction to male partners, female partners, or both." Thus the usual emphasis on the role of the environment has been replaced by "enduring attraction." Sexual orientation and gender identity are considered an immutable part of the self.[4]

While I fully support this view, why not let biology shine its light on *all* gender-related issues? This hate/love relationship is ideologically driven. Those who seek gender equality often find biology inconvenient. They believe that the easiest way to reach equality is by downplaying inborn sex differences. In contrast, in the fight against homophobia and transphobia biology is seen as a mighty ally. If we can prove a biological basis to homosexual conduct and transgender identity, this will silence those who claim them to be "unnatural" or "abnormal." Homosexual behavior in animals defangs these arguments.

I wish that we'd proceed the other way around, though. Instead of giving ideology precedence over science, we first need to get the science of gender in order. Ideally, we'd study this topic free from ideology. After this, we can worry about the societal goals that we have in mind and use whatever we have learned to work toward those goals. An amicus curiae brief to the U.S. Supreme Court in the *Lawrence v. Texas* case made the point that homosexual conduct is a normal aspect of human sexuality since it has been "documented in many different human cultures and historical eras, and in a wide variety of animal species." This 2003 case led to a landmark rejection of laws banning same-sex sex, sodomy, and oral sex between consenting adults. In another application of science, the discovery that gender identity is detectable in the brain has been used as an argument by transgender persons to obtain gender reassignments in their birth certificates and passports.[5]

To understand the evolution of homosexuality, we obviously need more evidence than the behavior of a few captive penguins. However, it is important to note that so far as we know, there are no "gay penguins." There is no evidence that some of these aquatic birds have an exclusive or even dominant orientation to their own sex. Silo and Roy, for example, didn't stay together. After six years, Silo left his mate and took up with Scrappy, a female from California. The breakup rocked the Manhattan gay scene. Many were disappointed, not least of all Rob Gramzay, the zoo's senior penguin keeper, who wistfully recalled that both males "seemed a good pair together."[6]

Among penguins, fluctuations in partnerships and partner sex are so common that it is better to consider these birds bisexual than homosexual. Moreover, these fluctuations are not limited to zoos, where they could be attributed to the occasional imbalance in the number of males versus females. A study of king penguins, in a colony of over one hundred thousand breeding pairs on the Antarctic Kerguelen Islands, found frequent homosexual displays, especially between males. The French ethologist Gwénaëlle Pincemy describes how both individuals "stretch their heads skyward and rotate the head back and forth in unison with eyes closed, taking 'peeks' at one another at the extreme end of the rotation." Whereas about a quarter of all displaying pairs were male-male, only very few went on to the next stage of bonding in which the partners recognize each other's calls. This allows a couple to come together after separation, which is crucial in a crowd of thousands of birds. But even though same-sex pairs rarely reached this bonded stage, the point is that some did, even in the wild.[7]

Nevertheless, the fascination with penguins and the politicization of their sex lives sometimes reaches idiotic depths. In 2019 the Sea Life London Aquarium upped the ante by adding gender assignment to the mix: it reported that two lesbian penguins were raising a gender-neutral chick. The chick, who had two moms, was claimed to be the first Gentoo penguin in history "not to be characterized as male or female." The aquarium's general manager went so far as to note that "it is completely natural for penguins to develop genderless identities as they grow into mature adults." This was news to any biologist! Apart from the fact that it's preferable to speak of a chick's sex rather than its gender, the aquarium provided no information on the individual's anatomy or self-evaluation of its sex. I'd love to have inspected the chick, but I'm pretty sure I'd have found an ordinary penguin chick whose sex was simply not revealed to the public.[8]

THE RHESUS MONKEY troops that I studied at the Henry Vilas Zoo went through yearly seasons of mating, pregnancy, and birth. These macaques are tough and don't mind the winter cold (their native habitat includes the Himalayas), but their sex life is arranged so that babies arrive all at once, with the first warmth of spring. To this end, the mating season starts in late September. That's when females hang out together and signal that they have sex on their minds. The males seem to need more time to get ready, but the females warm up for two months of mating by literally jumping on top of each other.

The most intriguing part of this sexual frenzy is that status differences among the females melt away. Rhesus monkeys are aggressive and abide by a strict hierarchy. During the mating season, however, females associate in the oddest combinations. They blithely ignore the distance created by rank disparities. The lowest-ranking female may ride on the back of the alpha female, whom she normally would carefully stay away from. What a sight! Mounting takes various forms, but in the most typical pattern, one female almost hangs on top of another with her body draped over her back. Females rarely perform the full-blown mating pattern of males, who mount their partners by firmly clasping their ankles with both feet. Known as the footclasp mount, the male stands a few inches above the ground while he thrusts vigorously. The absence of this mount in females doesn't mean that they aren't looking for sexual stimulation, though, because they often rub their genitals against each other's lower backs.[9]

The homosexual behavior of the closely related Japanese macaque has been documented extensively in the wild. In a park on the outskirts of Minoo city, in Japan, scientists discovered sexual consortships in an all-male troop. Two males would associate for a while, performing frequent footclasp mounts alternated with cuddling and grooming.[10] These bachelor males would probably soon disperse to join one of the larger mixed troops, where they'd mate with females. It is rare to find individuals with an overwhelming preference for their own sex, but it does happen. In our brown capuchin monkey colony at the Yerkes Primate Center, for example, Lonnie sought sex with other males so persistently that we considered him gay. Our diary described Lonnie's interaction with a same-age young male named Wicket:

Lonnie and Wicket began to court each other, then proceeded to mount. They couldn't decide who was going to be on top, but ended up taking turns. Finally, Lonnie went up to Wicket with his mouth hanging open and tongue dangling out. Wicket was sitting upright leaning back, letting Lonnie have access to his genitals. He let Lonnie do his thing for about a minute. Then he pushed him off of him. But Lonnie insisted and eventually they did it again about eight times.

Another male-male fellatio occurred during one of our studies at the Chimfunshi Wildlife Orphanage in Zambia. Jake Brooker, a British graduate student on the project, filmed an adolescent male chimpanzee who had been attacked. The victim, very upset and screaming after his lost fight, approached an adult male, who opened his mouth while gazing at the younger male's groin. The young male then inserted his penis into the other's mouth, leading to a perfunctory fellatio without ejaculation. This brief genital contact calmed him down.[11]

Same-sex sex in primates has been known for a long time. In 1949 the American ethologist Frank Beach noted that male monkeys regularly mount each other, sometimes achieving anal penetration, while ignoring nearby females. I once had the chance of discussing sexual behavior with Beach, who is considered the father of behavioral endocrinology. Coming from a generally gay-friendly country, where homosexual love has been legal for over two centuries, I was perplexed by its continued persecution in the United States. Beach shook his head at the moral pressure to demonize behavior that almost every animal on earth exhibits at least occasionally. He considered homosexual conduct a basic mammalian pattern.[12]

In an attempt at destigmatization, Beach waged a battle on multiple fronts. Together with an anthropologist, he reviewed sexual mores across the globe, showing that many cultures accept a wide range of practices. Their 1951 book *Patterns of Sexual Behavior* offered a comprehensive perspective that included cross-cultural data and extensive primate comparisons. The book was the first to drive a scientific nail into the coffin of the psychiatric position that homosexuality is a mental illness. Yet it took until 1987 before this "illness" was thrown out of the holy book of the American psychiatric world, the *Diagnostic and Statistical Manual of Mental Disorders* (DSM). After the horrible conversion therapies, lobotomies, and chemical castrations of not so long ago, this reclassification of homosexuality heralded a new approach. Nowadays the recommended treatment is gay-affirmative therapy, which seeks the acceptance of one's sexual orientation.[13]

Animal sexuality has played a remarkable role in the normalization of a love that dared not speak its name: it helped debunk the spurious argument that homosexuality defies the laws of nature. If heterosexuality is natural, so the thinking went, then homosexuality must be abnormal. As if there weren't room for both! This argument was finally put to bed in 1999 by a thick compendium of well-documented examples of same-sex sexual behavior in 450 different species. The Canadian biologist and linguist Bruce Bagemihl reviewed these cases in *Biological Exuberance: Animal Homosexuality and Natural Diversity*. He argued that reproduction is only one of the many functions of sex. Experts did not agree with every single description or interpretation given by Bagemihl, but his book did remove any doubt about the wide distribution of homosexual behavior in the animal kingdom.[14]

Bagemihl struggled to get his voice heard. Both scientists and laypeople tried to explain animal homosexual behavior away as something else, something asexual. It just couldn't be what it seemed to be. This tactic was also noted by the primatologist Linda Wolfe, one of the first to publish a field report on this behavior. Other researchers were skeptical of her observations, accusing her of doctoring photographs and making up stories about her monkeys. Wolfe complained, "They said that females were mounting each other by mistake—they didn't know what they were doing."[15]

Let's call this the confused-monkey hypothesis. Many other equally implausible ideas have been floated, such as the notion that homosexual behavior isn't truly sexual but rather represents "sham," "mock," or "pseudo" sex. Or it is just a way to express dominance (with the female role being submissive). Or that it never occurs voluntarily, or it is an artifact of captivity, or it occurs only when there is a surplus of either males or females, and so on. Some of these arguments contain a kernel of truth. When males spend a long time together without females, as in the all-male Japanese macaque troop mentioned above, sexual impulses often do find an outlet in homosexual behavior. This holds for both sexes and occurs also in our species, such as among sailors on a ship or nuns in a convent. But none of these counterarguments can explain the immense diversity of sexual behavior reviewed by Bagemihl, who summed up his frustration as follows:

> *When a male giraffe sniffs a female's rear end—without any mounting, erection, penetration, or ejaculation—he is described as being sexually interested in her and his behavior is classified as primarily, if not exclusively, sexual. Yet when a male giraffe sniffs another male's genitals, mounts him with an erect penis, and ejaculates—then he's engaging in "aggressive" or "dominance" behavior, and his actions are considered to be, at most, only secondarily or superficially sexual.*[16]

IT WAS A typical situation over dinner in a Rome restaurant: one human male was challenging another in front of his girlfriend. Knowing the gist of my writings, the man sought to provoke me by demanding, "Name one area in which it's hard to tell humans and animals apart!"

Before I knew it, between two bites of delicious pasta, I said, "The sex act."

This took him aback a little, but only momentarily. He launched into a great defense of passion as peculiarly human, the recent origin of romantic love, the poems and serenades that come with it, while pooh-poohing the anatomy of *l'amore,* which is the same for humans, hamsters, and guppies (guppy males have a penis-like modified fin.) He pulled a disgusted face at these mundane mechanics.

When his girlfriend, a colleague of mine, jumped in with more examples of animal sex, however, we had the sort of dinner conversation that delights primatologists but embarrasses most everyone else.

People always look at animal sexuality in purely functional terms, calling it "breeding behavior." There is no fun, no love, no gratification, and no variation, and it can occur only between a mature male and a fertile female. Perhaps we project onto animals the sort of sex lives we believe we *ought to* lead. Sex has only one purpose, so why use it for anything else? This is why our long list of sexual sins includes onanism, homosexuality, anal sex, and even birth control. Since we humans stray all the time from the morally sanctioned path and feel perhaps guilty about it, we double down on animals by insisting that they dedicate themselves purely to making babies. Never mind that in some species, such as the bonobo, three-quarters of sexual activity has nothing to do with reproduction. Sex occurs in combinations that simply cannot reproduce, or it takes forms that won't bring any sperm close to an ovum.

Bonobos are known as the hippies of the primate world. We find a "Bonobo Bar" in many large cities and hear of sex therapies that promise to "release your inner bonobo." These apes have become the favorites of the LGBTQ community even though I have yet to meet a predominantly homosexual one. Human categories don't apply to bonobos. On Alfred Kinsey's famous 0–6 rating scale from exclusively heterosexual to exclusively homosexual, most humans may be at the heterosexual end, but every bonobo is totally bi, or a perfect Kinsey 3.

Apart from male-female copulation in a great variety of positions, the most characteristic pattern is genito-genital rubbing between females. This belly-to-belly posture—in which one female may be lifted off the ground by the other while she clings to her much as an infant clings to its mother—allows both females to make rapid sideways movements. They rub their engorged clitorises together with an average of 2.2 lateral moves per second, which is the same rhythm as a thrusting male. Every student of bonobo behavior has observed GG rubbing, whether the bonobos are captive or in the wild.[17]

Bonobos show other postures and patterns that bring pleasure without a chance at fertilization. For example, males engage in rump-rump contact, in which both partners stand on all fours while briefly rubbing their butts and scrota together. Penis fencing, which has thus far been seen only at the Wamba field site, occurs when two males face each other, hanging from a branch while rubbing their penises together as if crossing rapiers.[18]

A regular erotic pattern is the open mouth kiss, in which one partner places his or her mouth over that of the other, often with extensive tongue-tongue contact. While typical of bonobos, such "French kissing" is absent in chimpanzees and most other primates, which perform more platonic kisses. This explains why a zookeeper unfamiliar with bonobos once accepted a kiss. Was he taken aback when he suddenly felt the ape's tongue in his mouth!

Males will stimulate each other's genitals manually. One male with his back straight and legs apart presents his erect penis, while the other loosely closes his hand around the shaft, making caressing up-and-down movements. The massage doesn't typically lead to ejaculation. Females also touch or poke each other's genitals but switch to GG rubbing as soon as there is more sexual interest between them. They prefer a more symmetrical interaction.

Because sex partners often face each other up close, facial expressions and sounds make their exchanges intense and intimate. We know from detailed video analyses by the Italian primatologist Elisabetta Palagi that partners have lots of eye contact, coordination, and synchronization. Females squeal loudly during GG rubbing, and if one partner excitedly bares her teeth, the other instantly mirrors her expression. Facial mimicry is a measure of empathy, as it is in humans, and it is more common between females than during heterosexual sex. Fieldworkers have also measured an increase in oxytocin in female urine following homosexual but not heterosexual sex. The implication is that females are emotionally more affected by contact with their own sex than with the other.[19]

These findings run counter to the idea that heterosexual contact is the pinnacle of sexual activity. That females preferentially seek each other out for sex and are emotionally more invested in it than in mating with a male fits the structure of a society run by a tight sisterhood. Female bonobos need to resolve conflicts and foster cooperation. Sex is their social glue.[20]

Lest this quick overview leave the impression that bonobos are an oversexed species, I must add that their erotic activity is utterly casual and relaxed. Due to our human obsession, we may find this hard to grasp. We have so many taboos and we so assiduously keep certain body

Bonobo females smooth their relationships through frequent GG rubbing. One female clings to another while they laterally rub their clitorises together in a rapid rhythm.

parts out of sight that we can't imagine ourselves without this mental straitjacket. We're never totally relaxed about sex. We censor nudity, measure skirt length at school, suppress sexual thoughts, and use a rich array of euphemisms to avoid mentioning sex or bodily functions altogether. Even accidentally brushing against a stranger's breast, buttocks, or genitals may be taken the wrong way. Sexuality is a forbidden fruit that we guard with a devotion and indignation that would be ridiculous in any other domain.

For bonobos, on the other hand, the fruit hangs low, and they pluck it whenever they feel like it. We can't call them "liberated" since they were never repressed to begin with. Our kinds of inhibitions and fixations are alien to them. Sex is no big deal for them. It's such a natural and spontaneous part of their lives that it is hard to detect a borderline between social and sexual affairs.

Although bonobos are a serious contender for the title "sexiest primate alive," it's not as if they don't do anything else. Like people, they have sex occasionally, not continuously. They initiate sex multiple times per day but don't do it all day. Most of their contacts, particularly those with or among youngsters, are not carried through to the point of sexual climax. Partners merely pet and fondle each other while focusing on the genitals. The average copulation between adults is quick by human standards, typically under fifteen seconds. Instead of an endless orgy, we see a social life peppered by brief moments of sexual enjoyment. In the same way that we shake hands or pat each other's backs, bonobos have "genital handshakes" to build relationships and signal good intentions.

THE EARLY 1990S saw the first reports about differences between male and female brains as well as the possibility of a "gay brain." In the Netherlands, these discoveries caused extraordinary turmoil. One gay leader claimed that linking sexual orientation to the brain risked turning homosexuality into a medical issue. Dick Swaab, the neuroscientist at the furor's center, was likened to Dr. Mengele, the Nazi doctor who had experimented on live prisoners. The paranoia ran so high that Swaab was subjected to anonymous death threats and bomb scares.[21]

Resistance to biology was part of a broader fear among social reformers on both sides of the Atlantic. They worried that references to brains and genes might thwart their ambition to change society. "Genes hold culture on a leash" was the provocative way the American sociobiologist and entomologist E. O. Wilson had put it. It didn't matter that Wilson reassured us that the leash was "very long." He, too, was branded a fascist.[22]

Nowadays, we look at the role of biology differently, certainly with respect to gender identity and sexual orientation. The LGBTQ community realizes that characterizing their lives as a mere "choice," "preference," or "lifestyle" crumbles in the face of the evidence for genetic, neurological, or hormonal factors. It takes the choice element out of it. But even though research in this area is no longer met with the same hostility as before, it's hardly free from controversy. When the American neuroscientist Simon LeVay pinpointed a specific brain area as a marker of sexual orientation, he was roundly criticized for making too binary a distinction between hetero- and homosexual men. Didn't these groups overlap in real life, and didn't they overlap in his own data? Among heterosexuals, a tiny area of the hypothalamus is on average twice as large in men as in women. Gay men, in contrast, have an area similar in size to that of women. Was LeVay suggesting that gay men constitute an intermediate gender of "effeminate" men? Even though himself openly gay, LeVay was accused of "seriously simplifying sexuality" and misrepresenting the "range of human possibility."[23]

It is unclear if this small brain area (the size of a grain of rice) contains all the answers. Follow-up studies have cast doubt on the earlier findings. On top of that, we have the usual chicken-and-egg problem of whether it's the brain that steers behavior in a given direction or the other way around. Couldn't the neural tissues in LeVay's study *reflect* the sort of lives that their deceased owners had led?[24]

It took almost two decades before Ivanka Savic and Per Lindström resolved this conundrum at the Stockholm Brain Institute in Sweden. Instead of inspecting the same brain area as LeVay, they focused on much more general neural traits, such as brain asymmetry, that have no direct relation to particular behavior. These brain features are fixed at birth and don't change with experience. Nevertheless, they reflect gender and sexual orientation. Brains of gay men are structurally similar to those of heterosexual women, whereas those of lesbian women resemble those of heterosexual men. Savic concluded that "these differences are likely to have been forged in the womb or in early infancy."[25]

Sexual orientation also may dictate the response to a chemical compound, androstadienone, that is secreted in male underarm sweat and bottled in aftershaves and hair gels. Even though we underestimate the power of our noses, odors steer us toward individuals with romantic potential. Inhalation of androstadienone has little effect on heterosexual men, but in heterosexual women and homosexual men, it triggers the hypothalamus. When asked, subjects say they don't find the stuff particularly attractive, but as so often with pheromones, androstadienone works unconsciously.[26]

In an unexpected "ramification" of the above research, similar distinctions were found in domestic sheep. Some healthy rams fail to mount estrus ewes; they used to be called "nonworking," "asexual," or "inhibited." But the American neuroendocrinologist Charles Roselli now considers that a mischaracterization. About one in twelve rams has a strong same-sex sexual preference. Far from being asexual, these individuals are eager to mount members of their own sex while ignoring nearby ewes. It is a stable individual trait. *Ovis aries* is only the second mammal, after ourselves, in which an exclusive homosexual orientation has been found.

Similar observations have been made in wild bighorn and thinhorn sheep. It is even said that ewes sometimes try to spark an adult ram's sexual interest by mimicking a young male's behavior. Rams may lick each other's penises, mutually rub, nibble, and nuzzle each other, and achieve anal penetration, pelvic thrusting, and ejaculation. As in us, their sexual orientation seems to be reflected in the hypothalamus, which contains a nucleus that is larger in female-oriented rams than in ewes. In male-oriented rams, on the other hand, its size falls somewhere in between.[27]

In short, even though the brain cannot tell us with certainty what sexual orientation an individual has, it does seem to contain a few markers. Like gender identity, sexual orientation appears to be present at birth or to develop soon thereafter. It is therefore part and parcel of who we are. This applies not only to members of the LGBTQ community but to all humans (and perhaps also sheep). Gender identity in general and sexual orientation in general are inalienable, inalterable aspects of every person.

This still doesn't mean that the situation is simple, though. For one thing, these findings fail to tell us where sexual orientation comes from. We have evidence for genetic factors, but not for a single gay gene—not even a handful of gay genes. The genes involved are multifold and distributed. It has long been known that being gay or lesbian runs in families, and that identical twins share a sexual orientation more often than nonidentical twins or other siblings. But this can't be the whole story. After all, identical twins also often differ. Despite carrying the same DNA, one of them can be gay and the other straight. The largest twin study to date included nearly four thousand pairs in the Swedish Twin Registry. The conclusion was that sexual orientation depends on a combination of familial and environmental effects. A person's genome alone can't tell us their sexual orientation.[28]

A second problem is a dichotomy underlying most studies. To divide sexual orientation into just two categories seems a gross simplification that overlooks the realities of human behavior. It is often thought that while women occupy the whole spectrum of orientation, men flock to the extremes. They are either attracted to their own gender or to the other one, never both. Bisexual people face discrimination from heterosexual and homosexual corners alike. Can't they make up their minds? Are they perhaps overly promiscuous? *Do you have a lot of threesomes?*

people ask them. For the longest time, science dismissed bisexuality as either a phase or a form of experimentation.

So much skepticism surrounded bisexuality that Kinsey, the American sexology pioneer, designed his 0–6 scale to show that the middle category also exists for men. Kinsey himself identified as bisexual. A recent reanalysis of previous studies confirms that self-declared bisexual men truly are interested in both genders. It's neither a ruse nor a phase. Measures of penile erections showed that they get aroused by watching erotic videos regardless of whether they feature a man or a woman. Maybe science will finally believe what bisexual people have been saying all along.

Kinsey, who was both famous and infamous for pointing out the immense gap between how we'd like humans to behave and our actual sexual practices, noted that homosexual and heterosexual orientations aren't easily pried apart. Many men declare themselves to be one or the other but are in fact both. The majority of men seems "mostly heterosexual" rather than exclusively heterosexual.[29] Ironically, before we knew much about the sex lives of sheep and goats, Kinsey delivered the following famous admonition:

> *Males do not represent two discrete populations, heterosexual and homosexual. The world is not to be divided into sheep and goats. Not all things are black nor all things white. It is a fundamental of taxonomy that nature rarely deals with discrete categories. Only the human mind invents categories and tries to force facts into separated pigeon-holes. The living world is a continuum in each and every one of its aspects. The sooner we learn this concerning human sexual behavior the sooner we shall reach a sound understanding of the realities of sex.[30]*

KINSEY WAS RIGHT about the human mind. We are a symbolic species, which means that we have a word for everything. Language makes us slice and dice the world into neat categories while closing our eyes to any possible blending. It is the exact opposite of how nature operates. As the American reproductive biologist Milton Diamond is fond of saying, "Nature loves variety. Unfortunately, society hates it."[31]

I have often thought about this issue in the racial context. We divide human races into black, white, brown, and yellow, while ignoring the enormous genetic variability and overlap underneath those skin colors. Genetically, the races are hard to tell apart, and every human carries a mixture of genes that comes from far away and long ago. There are no purebloods in the world even though we try to squeeze everyone into one or the other category.[32] We have a label for every race, and we sometimes call them names or act as if one were superior to the other. I have never noticed anything remotely similar in animals, however, despite the color variation common in many species. If an individual's appearance is radically different, alarm or hostility may follow, such as toward an albino newborn or individuals deformed by disease. But less dramatic variations arouse little reaction. For example, almost all chimpanzees and bonobos are black. Still, whenever a rare light brown individual is born, we never see any special treatment, neither positive nor negative.

Brown spider monkeys are another good example. This primate is also known as the *variegated* spider monkey due to its color variation from dark brown, almost black, to tawny and light tan. I have seen all color variants happily mixed together in captivity. But perhaps this isn't so remarkable given that they had no choice. So I asked Andrés Link about it—a Colombian colleague who studies these particular monkeys in the wild, where color variants mix as well. He told me that he's never noticed any behavioral biases toward lighter or darker variants, not even after the birth of two leucistic individuals in his population. Even though they are not albinos, these monkeys are as white as can be except for a few color patches and dark eyes. In Link's words, "Also these individuals are completely normal in their interactions with other group members."

Humans are different. As we do for the races, we have a plethora of labels for gender characteristics and sexual appetites. We often use these labels to communicate approval or disapproval. A desperately sad and extreme case of labeling was the pink triangle that the Nazis put on gay prisoners, whom they subjected to an extra level of cruelty. Despite this dark background, these triangles have recently popped up at gay pride events as badges of honor. More generally, we linguistically divide humanity's rich sexuality into what Kinsey called "sheep and goats," carving up what is essentially a continuum into just two or three categories. This is not to say that labeling is the *source* of transphobia or homophobia, because labels can also be applied in a more tolerant fashion. Many languages have a word (and room in their society) for a third gender. It remains true, however, that labels hand a powerful weapon to those who are phobic. Labels easily move from being descriptive to being hurtful and insulting. Being a symbolic species has its pros, but it comes with terrible cons.

I am not sure that *phobia* (extreme or irrational fear) is the right word for human sexual prejudice. While it is very possible that fear, insecurity, and suppressed sexual impulses lurk underneath the intolerance, other, more hostile emotions seem to be at play as well. Whatever they are, though, they aren't observed in other primates. Despite all the striking parallels between human and animal sexuality, the one aspect that's never reported is rejection based on sexual orientation or expression. Remember how Donna, the gender-nonconforming chimpanzee, was extremely well integrated. So was Lonnie, our "gay" capuchin monkey. Among the primates, I can envision rejection only for individuals who disturb the peace or otherwise interfere with others' lives, which is rarely the way homosexual tendencies are expressed. In fact, you'd think the opposite. Evolutionarily speaking, the hate that heterosexual men direct at gays is "deeply incomprehensible," as LeVay notes. Instead of objecting to males who have a different sexual taste, heterosexual men should be thrilled to see others waste their seed on each other instead of entering the fray over women.[33]

But even if we have all sorts of labels and social pressures to push people, especially men, to choose between hetero- and homosexuality, it's good to realize how recent a phenomenon this is. The term *homo-sexual* didn't exist until the nineteenth century. Before that time, there was plenty of homosexual conduct but no homosexual identity. Among men, same-sex sex was typically age-structured, with older men pene-trating younger ones, such as the soldiers of ancient Greece who boosted their bravery before setting off to war. During certain eras, sodomy was nearly universal, whereas lesbian relationships stayed mostly under the radar but were probably equally prevalent. In 1869 Karl-Maria Kert-beny, a German-Hungarian author, coined the twin terms *homosexual* and *heterosexual* to replace the pejorative labels that he despised. Since then, at least in the West, language began to promote a dichotomy that was unknown before. Homosexual activities used to be supplemental to heterosexual ones, often performed by men and women who at the same time were heterosexually married and had families. This may still be the case, but it is now obscured by the labeling that we have grown used to.[34]

The nonexclusivity of sexual orientation is important because if there is one issue that biologists have debated to death, it is how homosexual-ity could possibly have arisen. Some call it an evolutionary puzzle; others say it shouldn't exist. For example, in *Ever Since Adam and Eve: The Evo-lution of Human Sexuality,* Malcolm Potts and Roger Short bluntly state that "homosexual behaviour is the antithesis of reproductive success."[35]

This may sound logical, but it isn't if exclusive sexual orientations are rare. Reproduction is hardly in danger if some individuals seek sex within their own gender. Many people who call themselves lesbian or gay have brought children into the world at some point in their lives. Mathematical modeling of genetic traits has shown that homosexual orientations can easily emerge in a population. According to these mod-els, they should be quite common, and perhaps they are.[36]

Let me therefore rephrase the above question. To wonder how homo-sexual behavior could have evolved is the wrong approach. It buys into a doubtful dichotomy unsupported by what we know about genetics as well as actual human behavior. To my mind, the better question is whether we should be surprised that humans and other animals regularly engage in sexual activities that can't possibly lead to reproduction. Does evolutionary theory allow for such an opening up of sexual possibilities?

Of course it does. The animal kingdom is chock full of traits that evolved for one reason but are also used for others. The hooves of ungulates are adapted to run on hard surfaces, but they also deliver a mean kick to pursuers. The primate hand evolved to grasp branches, but it also allows infants to cling to their mothers, which is a smart thing to do high up in the trees. The mouths of fish are for feeding, but they also serve as holding pens for the fry of mouth-breeding cichlids. Color vision is thought to have come about because our fruit-picking primate ancestors needed to judge the ripeness of their food. But once we perceived color, this capacity became available for reading maps, noticing someone's blushing, or finding shoes that match our blouse.[37]

If bodies and senses often are multipurpose, the same applies to behavior. Its original function doesn't always tell us how it will be used in daily life, because behavior enjoys *motivational autonomy*.

THE MOTIVATION BEHIND behavior rarely includes the goals for which it evolved. These goals stay behind the veil of evolution. We evolved nurturant tendencies, for example, to raise our own biological children, but a cute puppy triggers these tendencies just as well. Whereas reproduction is the evolutionary goal of nurturance, it isn't part of its motivation.

After a mother dies, other adult primates often take care of her weaned juvenile. Humans, too, adopt on a large scale, often going through hellish bureaucratic procedures to add children to their families. Stranger yet is cross-species adoption, such as by Pea, a rescued ostrich at the David Sheldrick Wildlife Trust in Kenya. Pea was beloved by all orphaned elephant calves at the trust and took special care of a baby named Jotto, who'd stay by her side and sleep with his head on her soft feathered body. The maternal instinct is remarkably generous.[38]

Some biological purists call such behavior a "mistake." If adaptive goals are the measure, Pea was making a colossal error. As soon as we move from biology to psychology, however, the perspective changes. Our impulse to take care of vulnerable young is real and overwhelming even outside the family. Similarly, when human volunteers push a stranded whale back into the ocean, they employ empathic impulses that, I can assure you, didn't evolve to take care of marine mammals. Human empathy arose for the sake of family and friends. But once a capacity exists, it takes on a life of its own. Rather than calling the saving of a whale a mistake, we should be glad that empathy isn't tied down by what evolution intended it for. This is what makes our behavior as rich as it is.

This line of thought can also be applied to sex. Even if our genital anatomy and sexual urges arose to bring about fertilization, most of us engage in sex without paying much heed to its consequences. I've always thought that the main impetus for sex must be pleasure, but in a poll conducted by the American psychologists Cindy Meston and David Buss, people offered a bewildering array of reasons for engaging in it. They ranged from "I wanted to please my boyfriend" to "We had nothing else to do" and "I was curious how she'd be in bed." If humans usually don't think about fertilization while making love, animals, who don't know the connection, do so even less. At least, I've never seen evidence that they do. They have sex because they are attracted to each other or have learned that it gives them pleasure, not because they want to reproduce. One can't want something one is unaware of.[39]

Motivational autonomy allows the sex drive to apply outside fertile gender combinations. It is free to team up with that other reality of social life: same-gender bonding. In all primates, young males seek out males and young females seek out females as playmates, thus creating sex-segregated social spheres that last into adulthood. These spheres provide great satisfaction and enjoyment, which occasionally spills over into sexuality. The sharp boundary in human society between the social and sexual domains is artificial. It's a cultural invention that, despite moral and religious exhortations, is prone to leakage.

Looked at this way, homosexual behavior is nothing out of the ordinary. When Joan Roughgarden reviewed existing theories about the evolution of homosexuality, she settled on gratification through physical intimacy in mammals "who happen to have genitals filled with pleasure sensing neurons and who happen to use their genitals for signaling and social purposes other than the exchange of gametes in heterosexual mating." This may still be the best way to look at same-sex sex: not as a specifically evolved trait that starkly contrasts with heterosexual behavior, but as the outcome of powerful sexual urges and pleasure-seeking tendencies mixed with same-gender attraction.[40]

Despite the many parallels between same-sex sex in humans and other animals, the big difference is our tendency to carve up and label sexual behavior and orientations. The labeling sets us up for intolerance. I love how other primates take every individual as they come without hang-ups about whether they conform to the majority.

13

THE TROUBLE WITH DUALISM

Mind, Brain, and Body Are One

Having a second child is the perfect antidote to the illusion that we control the way our children turn out. Parents may have viewed their firstborn as putty in their hands, ready to be molded this way or that. But with the second child, while pursuing the same parenting style, they inevitably get a different outcome. As Mary Midgley put it in her dedication to *Beast and Man,* "To my sons, with many thanks for making it so clear that the human infant is not blank paper."[1]

This realization is amplified when parents get a daughter after a son or a son after a daughter. Now it's about more than individual temperament: gender comes into play. It is a rare parent who, after this experience, still elevates nurture over nature.

Yet in academic discourse, nurture still often is the sole message. I'm at a loss as to why this is so, and I've tried to poke holes into this position by describing how males and females behave in our next of kin. Not that the conclusion is crystal clear, but at a minimum, it's considerably richer than the cliché of the male monkey overlord that was foisted upon us during an era of limited knowledge. Human gender differences have enough commonalities with primate sex differences to clarify that we didn't escape the forces of evolution.

And I haven't even touched on the role of hormones and the brain, which would have added a whole other biological dimension. I'm reluctant to do so because I'm no specialist in these areas, although I've been surrounded all my life by colleagues who are. My exposure to their work has taught me that nothing is simple. Even a routine-sounding statement, such as that testosterone drives violence, is tricky. We look at this hormone as the essence of manliness and say of a cocky guy that he must be brimming with testosterone. We shouldn't be putting all the blame on his raging hormones, however. First off, women produce testosterone as well, although at lower levels. And while aggressive behavior does require testosterone, which is why castration can bring it down, there is no simple one-on-one relation. Whenever male monkeys are put together in captivity, their testosterone levels fail to predict which one will be the most aggressive. On the contrary, the amount of aggression that each individual shows predicts his subsequent hormone level. Hormones and behavior affect each other mutually.[2]

Regarding the brain, we run into a similar problem. Are the brains of men and women distinct from birth, or do they grow apart due to different societal pressures? In *The Gendered Brain,* the British neuroscientist Gina Rippon defends the latter position, attributing sex differences in the brain to life's experiences. She claims that the human brain starts out being as gender-neutral as the liver or the heart.[3] Neither livers nor hearts are gender neutral, though, and other neuroscientists argue that brains are masculinized or feminized under the influence of hormones in the womb. Rippon's book has been criticized for downplaying well-known brain differences. The British psychologist Simon Baron-Cohen, for example, believes that autism spectrum disorder (which is three to four times more common in boys than in girls) is an extreme expression of the typical male brain.[4]

The debate is complex and heated, not least because of the flying accusations of "neurosexism." The only point both sides agree on is that the brains of women and men are more similar than they are different.

Animals play a critical role in this debate, as their brains develop independently from the human cultural environment. If their brains vary by sex—and they do—why should ours be gender neutral? A recent study of capuchin monkeys, for example, reported a remarkable divergence between male and female brains, which differ in cortical areas associated with higher-order functioning. These areas are more elaborate in females than males. But here, too, we can't rule out experience-based brain changes, given how different the lives of male and female capuchins are.[5]

THESE CONTROVERSIES HAVE generated more than twenty thousand scientific articles on sex differences in the brain. I'll gladly leave it to the experts to decide how substantial they are. My goal in this book has instead been to compare human behavior with that of other primates.

I realize that many people prefer to keep animals at arm's length. In Rippon's words: "Not those bloody monkeys again!" Outside science and sometimes within, a common belief seems to be that, while our bodies are a product of evolution, our minds are ours alone. Humans aren't subject to the same laws of nature as animals, and we feel and think the way we do because we freely chose to. I consider this position a form of neo-creationism: it neither denies nor fully embraces evolution. As if evolution came to a screeching halt when it reached the human (and only the human) neck, thus leaving our lofty heads alone!

It's all vanity. While our species is blessed with language and a few other intellectual advantages, socio-emotionally we are primates through and through. We are equipped with a large monkey brain and the psychology that it entails, including how we navigate a world of (mainly) two sexes. Calling them "genders" doesn't change all that much. However refined our rhetoric may get, it can never fully disentangle the cultural category of "gender" from the biological one of "sex" and the bodies, genitals, brains, and hormones that come with it. It's a bit like medieval nobility calling themselves blue-blooded, whereas we all know that as soon as a lance hits them, they will still be spilling red. Basic human biology shines through.

That we are gendered by nature doesn't diminish the value of the gender concept, however. Insofar as it draws attention to the cultural overlays, the learned roles, and the expectations society imposes on each sex, it is a powerful addition to the discussion. The juxtaposition between gender and sex makes the point that there are always two influences on everything we do: biology and the environment. We can't discuss the differences between men and women without considering both influences. This is also why it's instructive to explore sex differences within the triangular framework central to this book—humans, chimpanzees, and bonobos—because comparing ourselves to other primates adds the role of evolution.

The picture that emerges is anything but straightforward, however. The problem, if we call it that, is the variation among these three hominids. Our two closest relatives are quite different characters. Chimpanzees are far more bellicose than bonobos, and each features a radically different dynamic between the sexes. This fact alone precludes a simple evolutionary scenario even though some scientists try to arrive at one by kicking bonobos to the curb, dismissing them as the black sheep of our family. Being an observer by nature, I notice that whenever bonobos come up in a discussion, my colleagues often shift uneasily in their seats, smirk in embarrassment, scratch their heads, and generally display discomfort. Bonobos are highly inconvenient to those who build evolutionary narratives around male specialties, such as hunting and warfare. Chimpanzees fit these ideas so much better. Our current knowledge of genetics and anatomy, however, offers no reason whatsoever to favor chimpanzees over bonobos as models of our shared ancestor.

The mosaic of differences among these three hominids can't hide a few universal traits, though. Males are more status-oriented, and females are more oriented toward vulnerable young. Males are physically (if not always socially) dominant and more inclined to overt confrontation and violence, whereas females are more nurturant and dedicated to progeny. These tendencies manifest themselves early in life, such as in the high energy level and roughhousing of young males and in the attraction to dolls, infants, and baby-sitting of young females. This archetypical sex difference marks most mammals, from rats to dogs and from elephants to whales. It evolved thanks to the distinct ways the sexes transmit their genes to the next generation.

But not even this pronounced sex difference is absolute. It follows the usual bimodal distribution with overlapping areas and room for exceptions. Within each species, not all males and females are the same, and the differences we see are descriptive, not prescriptive. No one says that males *ought* to act this way and females that way, only that the sexes generally follow different agendas that lead them to behave differently from each other.

Other proposed gender differences have proved hard to confirm. It is often said, for example, that males are more hierarchical and make better leaders, whereas females are more peace-loving. Females are also said to be more sociable and less sexually promiscuous than males. In all these domains, my explorations have turned up only minor differences or none at all. Competition among females, albeit less physical, is common and intense. Female sex lives seem every bit as adventurous as those of males. And both sexes arrange themselves in social hierarchies and enjoy lifelong friendships, even though the details differ.

And then there are the exceptions to the rule, which hint at flexibility in our behavior as well as that of our fellow hominids. Male apes can be remarkably nurturant, for example, and females can make great leaders. The latter is true not only for species with female dominance, such as the bonobo, but even for those with male dominance, such as the chimpanzee. If we look beyond males' physical strength advantage and focus on who decides group processes, both sexes demonstrate power and leadership.

The most exceptional social feature of the human primate is a family structure that ties men and women together. As a result, the genders are more interdependent in us than in our nearest relatives. The genders' integration is further amplified in modern society, where we ask them to work together not only in the family but also in the workplace. This is a significant departure from the role division of small-scale human societies. To welcome women into the public sphere and have them fully participate, however, will require a realignment of duties on the family front. Men will need to get more involved at home to get the genders' respective workloads back into balance. Our primate background may resist this shift, but perhaps a greater obstacle is the way our economies are structured. Traditionally, men gained a wage or salary working outside the home, whereas women gained none working inside. Even though biology has been invoked to justify this odd arrangement, really nothing in the nature of the human male prohibits him from taking on childcare, let alone other household tasks.

Our biology is more flexible than people think. The same flexibility marks our fellow hominids. This may sound surprising given the extent to which we have been indoctrinated to think of animals as preprogrammed machines. Animal behavior is still often attributed to instinct, whereas human behavior is viewed as a cultural product. This dichotomy is outdated, given what we have learned in the last few decades about animal cognition and behavior. It is particularly strange in relation to animals who nurse for at least four years and take almost as much time to reach maturity as we do.

No species should delay reproduction unless it's absolutely essential for survival. The only plausible reason for the slow development of apes is that their young require many years of learning and instruction to grow into competent adults. This is how humanity's prolonged immaturity is explained, and the same holds for other slowly developing species. Their societies are complex, and they need lots of knowledge and skills to be successful. There is no reason, therefore, to consider apes any more or less instinctive than we are.

Apes are products of their environment, too. They emulate, imitate, and adopt the habits of those around them. My team has conducted many studies of how apes learn from each other, and all I can say is that the English verb *to ape*, and its equivalent in other languages, is most felicitous. Apes have a talent for watching and learning. Like children, young apes seek out adult models of the sex they identify with. Females typically copy their mothers, whereas males follow high-status males. As a result, both sexes learn sex-typical behavior partly from their elders.[6]

This makes apes gendered, too.

WESTERN RELIGION AND philosophy have traditionally defined us in opposition to nature instead of in line with it. Since we like to place ourselves above the beasts and close to the angels, we almost resent our bodies. They remind us too much of our lowly origins and bother us every day with uncontrollable lusts, needs, ailments, and feelings. How did the magnificent human spirit end up trapped in such a flawed material vessel? As the Gospel of Thomas laments, "I am amazed at how this great wealth has made its home in this poverty."[7]

The mind is godly, the body not so much. This dualism is quintessentially masculine, concerned less with the human mind than with the male mind. It's always men who have tried to convince themselves that their intellect floats on a plane high above biology. This stance is easier to maintain if your body doesn't go through hormonal cycles. Moreover women's bodies bleed, which men have traditionally portrayed as disgusting and "impure." Through the ages, men have sought to distance themselves from the flesh (weak), emotions (irrational), women (childish), and animals (dumb).

Given that men are just as closely tied to their bodies as are women and animals, these contrasts are entirely illusory. They are figments of the masculine imagination. Mind, brain, and body are one. There exists no nonmaterial mind. "No body, never mind," wrote the Portuguese-American neuroscientist Antonio Damasio. "A mind is so closely shaped by the body and destined to serve it that only one mind could possibly arise in it."[8]

What is most puzzling is that modern feminism has embraced the same age-old dualism, featuring a familiar denial of the body. The human infant, in this view, is born genderless, with a gender-neutral brain that awaits instruction from its environment. We are what we want to be, or at least what society wants us to be, without much input from the vessel that carries us around. The vessel walks, talks, eats, poops, reproduces, and performs other mundane survival tasks, but its gender is up to the mind.

Mind-body dualism is an eternal philosophical topic about which much more has been written than I could ever read. My chief interest has always been its application to animals and the offensive Cartesian notion that they lack souls, but let me briefly (and no doubt superficially) mention dualism in relation to gender. The idea can be traced back at least to Plato, and probably further. While the Greek philosopher's *Republic* famously affirmed equality between men and women, the Platonic *dialogues* are sprinkled with male chauvinistic remarks. The body is seen as an annoying obstacle. It is compared to a tomb or prison, and those who pay too much attention to it fail to do justice to their souls. Women exemplify this imbalance by being too close to their bodies while getting carried away by the emotions these call forth. Since women allow their bodies to compromise their souls, they lack the capacity for complete wisdom. Plato exhorts men to avoid living "womanish" lives.[9]

The same contempt for the body explains why medieval hermits—overwhelmingly male—sought to negate it. They'd withdraw to the desert or a nearby cave to deprive themselves of all temptations of the flesh—only to be tormented by visions of lavish meals and voluptuous women. It is also why rich people—again, almost exclusively men—line up to have their brains cryogenically frozen after death. They are so sure that the mind can do without a body that they pay a fortune for a digitally immortal future, which will be achieved when everything that is currently in their head is "uploaded" to a machine.[10]

Giving the mind priority over the body was never popular among women until second-wave feminism came along after World War II. If bodies are the source of our denigration, these women seemed to have concluded, let's declare them irrelevant and, except for what's between the legs, identical to those of men. This tendency to get around the body and emphasize the mind instead may have waxed and waned over the years, and it's not unanimous in the feminist movement, yet it remains recognizable today.

In an insightful article entitled "Woman as Body," the American philosopher Elizabeth Spelman warned: "Some feminists have quite happily adopted both the soul/body distinction and relative value attached to soul and to body. But in doing so, they may be adopting a position inimical to what on a more conscious level they are arguing for."[11]

Spelman reviewed statements from prominent feminists of that period, including Simone de Beauvoir and Betty Friedan, who elevated mental activities over bodily ones. Women were urged to take up "higher" intellectual creativity so that they could join men in their realm of transcendence. Women's bodily functions, such as those related to childbirth, were dismissed as dreadful and barbaric. Motherhood wasn't held up as a strength. On the contrary, one feminist called pregnancy a "deformation" while suggesting that it would be nice if women one day could escape it. Spelman concluded that "what woman's liberation ultimately means is liberation from our bodies."[12]

Not all feminists view the emulation of men as the path to equality. Nowadays many of them embrace and celebrate the female body, its unique role in procreation, and the pleasure and empowerment that it affords. But still, dualism sneaks into the discourse every time sex differences are minimized or questioned, as they routinely are. The more radical the notion of gender as a social construct, the less room remains for the body.

I WOULD NEVER WANT to live in a genderless or sexless world. It would be an incredibly boring place. Imagine if everyone looked like me, millions of gray-haired old white men. Even if we included males of all ages and races, humanity would still remain greatly impoverished. I have nothing against men, and some of them are my best friends, but what makes life interesting, exciting, and emotionally satisfying is the variety of people we meet, work with, and live with, people with different languages, ethnicities, ages, and genders. The masculine and feminine versions of *Homo sapiens* complement each other, and for most of us, sexual attraction is intensely gendered.

At the very least, this mixture of women, men, and children from a variety of backgrounds makes life interesting. But I also believe that we derive great pleasure from it. This is why I am always taken aback by calls for a gender-neutral society, one in which biological sex hardly matters. The idea is that the world would be a better place without different sexes or at least with less attention to them. This goal is not only unrealistic but also misguided. It is telling that these calls rarely spell out what is wrong with having sexes or genders. The problem is not their existence but rather the associated prejudices and inequities as well as the limitations of the traditional binary, which leaves out some among us. Society doesn't recognize all gender manifestations, doesn't accept all sexual orientations, and fails to treat the genders as equals. These issues are formidable and undeniable, and I agree that we need to work on resolving them. But rather than blaming the age-old sex divide per se, we should address the deeper problem of social bias and injustice.

For anyone wishing to change these attitudes, a great start would be to move away from mind-body dualism. A doctrine upheld by a two-millennia-long slew of male thinkers to elevate their souls above the rest of creation, including women, is unlikely to be helpful in dismantling gender prejudices. Moreover mind-body dualism is out of touch with everything we have learned from modern psychology and neuroscience. The body, which includes the brain, is central to who and what we are. By running away from our bodies, we only run away from ourselves.

How deeply our bodies affect us is obvious from the latest research. Given how much gender identity and sexual orientation resist change, most neuroscientists believe they are anchored in the human brain. We have learned this from LGBTQ children, whose identities and orientations violate expectations. Society may discourage and punish these children all it wants, but it can't quell their inner convictions. These convictions come from inside their bodies, not from outside. The same applies to the heterosexual majority of people. Their sexual orientation and gender identity, too, are an immutable part of whom they are. Subjecting a boy to years of feminine socialization, as John Money tried, still can't make a girl out of him.

Evidently, the social environment doesn't hold all the cards. The limits of socialization are also clear from sex differences observed across the globe. Cultural universals reflect our species' biological background. This argument is amplified if the same differences also mark our fellow primates. It is hard to watch male and female apes interact and not notice the parallels with our own behavior.

Despite the evidence that nature sometimes trumps nurture, however, we have no need to choose between the two. The most productive approach is to consider both. Everything we do reflects the interplay between genes and environment. Since biology is only half of the equation, change is always within reach. Few human behaviors are rigidly preprogrammed. I'm a biologist, but I'm also a firm believer in the power of human culture. I have direct experience with how gender relations vary from country to country. Within bounds, they're subject to education, social pressure, custom, and example. And even the few aspects of gender that resist change and seem inalterable offer no excuse to deprive one gender of the same rights and opportunities as the other. I have no patience with notions of mental superiority or natural dominance between the genders, and I hope we'll leave those behind.

It all comes down to mutual love and respect and appreciation of the fact that humans don't need to be the same to be equal.

ACKNOWLEDGMENTS

My public lectures have taught me how thirsty people are for knowledge about gender biology. However casually or briefly I may allude to sex differences in primates, audiences zero in on them. They want to hear what these differences mean for human society. My answers are met by nods of agreement, surprised laughs, or skeptical frowns, but they leave no one cold.

Gender remains one of the most sensitive and controversial topics. It's an ideological minefield where one easily says something wrong or is misunderstood. No wonder most people hem and haw when asked to address this topic. Writing a whole book about it may turn out to have been one of my most foolish decisions.

In writing this one, I've largely restricted myself to my area of expertise, which is the social behavior of anthropoid apes and how it compares to that of our species. I have had no shortage of published studies to work with. In addition, I have been personally close to lots of individual primates and am grateful for what they have taught me. My book features their personalities and behavior to bring the topic alive. It seeks to dispel erroneous notions about our fellow primates while pointing out what I think their behavior means for the ongoing gender debates.

I have greatly benefited from feedback by colleagues who have read chapters or provided valuable information. These are not only fellow primatologists and co-workers but also experts on human psychology or biology in general. That many of them are women may have helped me circumvent the masculine bias that I inevitably bring to the table. But let me emphasize that the ultimate responsibility for any claims or opinions expressed in *Different* rests with me.

Readers and helpers were Andrés Link Ospina, Anthony Pellegrini, Barbara Smuts★, Christine Webb★, Claudine André, Darby Proctor★, Devyn Carter★, Dick Swaab★, Donna Maney, Elisabetta Palagi, Filippo Aureli, Joan Roughgarden★, John Mitani, Joyce Benenson, Kim Wallen, Laura Jones, Liesbeth Feikema, Lynn Fairbanks, Mariska Kret, Matthew Campbell, Melanie Killen, Patricia Gowaty, Robert Martin, Robert Sapolsky, Ruth Feldman, Sarah Brosnan★, Sarah Blaffer Hrdy★, Shinya Yamamoto, Takeshi Furuichi, Tim Eppley, Victoria Horner, and Zanna Clay. (I have marked readers of multiple chapters with a star.) In addition, I learned much from feedback on the entire manuscript from Bella Lacey and two millennial lay readers, Sydney Ahearn and Loeke de Waal. I thank all of them from the bottom of my heart.

I am grateful to the Royal Burgers' Zoo, the Yerkes National Primate Research Center, the Wisconsin National Primate Research Center, the San Diego Zoo, and the Lola ya Bonobo Sanctuary in the Democratic Republic of the Congo for opportunities to conduct research. I thank Toshisada Nishida for inviting me to the Mahale Mountains National Park in Tanzania, as well as Emory University and Utrecht University for providing the academic environment and infrastructure to make this kind of work possible. I thank Takumasa Yokoyama, Christine d'Hauthuille, Victoria Horner, Desmond Morris, and Kevin Lee for allowing me to include their photographs. I am immensely lucky to have Michelle Tessler as my agent and John Glusman as my editor at Norton.

Both have always believed in me, and both enthusiastically encouraged and supported this writing project.

In 2020, the COVID-19 crisis and self-quarantine provided me with an unplanned "writer's retreat" spent with my soulmate, Catherine Marin, in the comfort of our Georgia home. Both of us have always held academic jobs, but we now enjoy retirement. We managed to avoid the virus, and we also survived the tumultuous national election in which our state played a pivotal role, while regularly going on pleasant hikes through nearby Stone Mountain Park. Catherine has been the first and foremost critical reader of my daily output and has greatly helped me stylistically. Her love and support through our fifty years together have made (and still make) all the difference.

NOTES

INTRODUCTION

1. Jacob Shell (2019).
2. Kings 3:16–28; Agatha Christie (1933).
3. *APA Guidelines for Psychological Practice with Boys and Men* (American Psychological Association, 2018), p. 3; Pamela Paresky (2019).
4. Hegel, "The Family," in *Philosophy of Right* (1821), www.marxists.org/reference/archive/hegel/works/pr/prfamily.htm.
5. Mary Midgley in Gregory McElwain (2020), p. 108.
6. Charles Darwin to C. A. Kennard, January 9, 1882, Darwin Correspondence Project, darwinproject.ac.uk/letter/DCP-LETT-13607.xml.
7. Janet Shibley Hyde et al. (2008).
8. On Solly Zuckerman's baboon study, see Chapter 4.
9. Arnold Ludwig (2002), p. 9.
10. Patrik Lindenfors et al. (2007).
11. Packer quoted in Erin Biba (2019).
12. Frans de Waal (2019); Chapter 9.
13. Christophe Boesch et al. (2010); Chapter 11.
14. C. Shoard, Meryl Streep: "We hurt our boys by calling something 'toxic masculinity,'" *Guardian,* May 31, 2019.
15. Frans de Waal (1995).
16. David Attenborough narrates a night out in Banff, May 15, 2015, www.youtube.com/watch?v=HbxYvYxSSDA.

17. Judith Butler (1988), p. 522.

18. Vera Regitz-Zagrosek (2012); Larry Cahill, ed., An issue whose time has come: Sex/gender influences on nervous system function, *Journal of Neuroscience Research,* 95, nos. 1–2 (2017).

19. Robert Mayhew (2004), p. 56.

20. Jason Forman et al. (2019).

21. *NIH Policy on Sex as a Biological Variable,* n.d., https://orwh.od.nih.gov/sex-gender/nih-policy-sex-biological-variable; Rhonda Voskuhl and Sabra Klein (2019); Jean-François Lemaître et al. (2020).

22. Roy Baumeister et al. (2007).

CHAPTER 1. TOYS ARE US

1. Marilyn Matevia et al. (2002).

2. Roger Fouts (1997).

3. Judith Harris (1998), p. 219.

4. Gerianne Alexander and Melissa Hines (2002).

5. Janice Hassett et al. (2008).

6. Christina Williams and Kristen Pleil (2008).

7. Christina Hof Sommers (2012).

8. Patricia Turner and Judith Gervai (1995); Anders Nelson (2005).

9. Deborah Blum (1997), p. 145.

10. Sonya Kahlenberg and Richard Wrangham (2010). Also see the interview with Wrangham in Melissa Hogenboom and Pierangelo Pirak, The young chimpanzees that play with dolls, BBC, April 7, 2019, www.bbc.com/reel/playlist/a-fairer-world?vpid=p03rw3rw.

11. Tetsuro Matsuzawa (1997).

12. Carolyn Edwards (1993); Chapter 11.

13. Margaret Mead (2001, orig. 1949), pp. 97, 145–48.

14. Shalom Schwartz and Tammy Rubel (2005).

15. Margaret Mead (2001, orig. 1949), p. xxxi.

16. Jennifer Connellan et al. (2000); Svetlana Lutchmaya and Simon Baron-Cohen (2002).

17. Brenda Todd et al. (2018).

18. Vasanti Jadva et al. (2010); Jeanne Maglaty (2011).

19. Anthony Pellegrini (1989); Robert Fagen (1993); Pellegrini and Peter Smith (1998).

20. Jennifer Sauver et al. (2004).

21. Janet DiPietro (1981); Peter Lafreniere (2011).
22. Stewart Trost et al. (2002).
23. Maïté Verloigne et al. (2012).
24. Pedro Hallal et al. (2012).
25. Anthony Pellegrini (2010).
26. Eleanor Maccoby (1998).
27. Carol Martin and Richard Fabes (2001), p. 443.
28. U.S. Government Accountability Office, GAO-18-258, March 2018.
29. Marek Spinka et al. (2001).
30. Dieter Leyk et al. (2007).
31. Kevin MacDonald and Ross Parke (1986); Michael Lamb and David Oppenheim (1989), p. 13.
32. *Toledo Blade,* November 13, 1987; Anthony Volk (2009).
33. Rebecca Herman et al. (2003).
34. Lynn Fairbanks (1990).
35. Elizabeth Warren, April 25, 2019, twitter.com/ewarren.
36. Cathy Hayes (1951); Robert Mitchell (2002).

CHAPTER 2. GENDER

1. John Money et al. (1955).
2. The sexes: biological imperatives, *Time,* January 8, 1973, p. 34.
3. Milton Diamond and Keith Sigmundson (1997); John Colapinto (2000).
4. Heino Meyer-Bahlburg (2005).
5. Siegbert Merkle (1989); David Haig (2004); Robert Martin (2019); Caroline Barton, How to identify a puppy's gender, TheNest.com, n.d., pets. thenest.com/identify-puppys-gender-5254.html.
6. Gender and health, World Health Organization, www.who.int/health -topics/gender.
7. Elizabeth Wilson (1998).
8. Alice O'Toole et al. (1996) and (1998); Alessandro Cellerino et al. (2004).
9. Clayton Robarchek (1997); Douglas Fry (2006).
10. Nicky Staes et al. (2017).
11. Elizabeth Reynolds Losin et al. (2012).
12. Ronald Slaby and Karin Frey (1975), p. 854.
13. Carolyn Edwards (1993), p. 327.
14. William McGrew (1992); Elizabeth Lonsdorf et al. (2004); Stephanie Musgrave et al. (2020).

15. Beatrice Ehmann et al. (2021).

16. Suzan Perry (2009).

17. Frans de Waal (2001); Frans de Waal and Kristin Bonnie (2009).

18. Axelle Bono et al. (2018).

19. Aaron Sandel et al. (2020).

20. Ashley Montagu (1962) and (1973); Nadine Weidman (2019).

21. Melvin Konner (2015), p. 206.

22. Richard Lerner (1978).

23. Hans Kummer (1971), pp. 11–12.

24. Frans de Waal (1999); Carl Zimmer (2018).

25. Ronald Nadler et al. (1985).

26. Robert Martin (2019).

27. Anne Fausto-Sterling (1993).

28. Expert Q&A: Gender dysphoria, American Psychiatric Association, n.d., www.psychiatry.org/patients-families/gender-dysphoria/expert-q-and-a.

29. Rachel Alsop, Annette Fitzsimons, and Kathleen Lennon (2002), p. 86.

30. Andrew Flores et al. (2016).

31. Jan Morris (1974), p. 3.

32. Devon Price (2018).

33. Selin Gülgöz et al. (2019).

34. Selin Gülgöz et al. (2019), p. 24484.

35. Jiang-Ning Zhou et al. (1995); Alicia Garcia-Falgueras and Dick Swaab (2008); Swaab (2010); Between the (gender) lines: the science of transgender identity, *Science in the News,* October 25, 2016, sitn.hms.harvard.edu/flash/2016/gender-lines-science-transgender-identity.

36. Ai-Min Bao and Dick Swaab (2011); Melissa Hines (2011).

37. Joan Roughgarden (2017), p. 502.

CHAPTER 3. SIX BOYS

1. José Carreras, interview (2016), smarttalks.co/jose-carreras-pavarotti-was-a-good-friend-and-a-great-poker-player.

2. Tara Westover (2018), p. 43.

3. Martin Petr et al. (2019).

4. Nora Bouazzouni (2017).

5. Bonnie Spear (2002).

6. Nikolaus Troje (2002); video of human locomotion at Bio Motion Lab, n.d., www.biomotionlab.ca/Demos/BMLwalker.html.

7. Jeffrey Black (1996).

8. Ashley Montagu (1962); Melvin Konner (2015), p. 8.

9. Frans de Waal (2019).

10. Martha Nussbaum (2001).

11. Lisa Feldman Barrett et al. (1998); David Schmitt (2015); Terri Simpkin (2020).

12. Saba Safdar et al. (2009); Jessica Salerno and Liana Peter-Hagene (2015).

13. George Bernard Shaw (1894); Antonio Damasio (1999); Daniel Kahneman (2013).

14. Simone de Beauvoir (1973), p. 301; Judith Butler (1986); Elaine Stavro (1999).

15. Adolescent pregnancy and its outcomes across countries (fact sheet), Guttmacher Institute, August 2015, www.guttmacher.org/fact-sheet/adolescent -pregnancy-and-its-outcomes-across-countries.

16. On Dutch sexual education, see Saskia de Melker, The case for starting sex education in kindergarten, PBS, May 27, 2015, www.pbs.org/newshour/ health/spring-fever.

17. Belle Derks et al. (2018); World Bank open data, data.worldbank.org

18. Nathan McAlone (2015).

19. A Disney dress code chafes in the land of haute couture, *New York Times,* December 25, 1991.

20. Dutchman Ruud Lubbers in 2004; Frenchman Dominique Strauss-Kahn in 2011.

21. Public opinions about breastfeeding, Centers for Disease Control and Prevention, December 28, 2019, www.cdc.gov/breastfeeding/data/health styles_survey.

22. Tanya Smith et al. (2017).

23. James Flanagan (1989), p. 261.

24. Frans de Waal (1982); John Carlin (1995).

25. Dominic Mann (2017).

26. Frans de Waal, The surprising science of alpha males, TEDMED 2017, ted.com/talks/frans_de_waal_the_surprising_science_of_alpha_males.

27. Frans de Waal et al. (2008); Jorg Massen et al. (2010); Victoria Horner et al. (2011).

28. John Gray (1992).

CHAPTER 4. THE WRONG METAPHOR

1. Frans de Waal (1989); Ben Christopher (2016).

2. Solly Zuckerman (1932), p. 303.

3. Kenneth Oakley (1950).

4. Jan van Hooff (2019), p. 77.

5. Lord Zuckerman (1991).

6. Richard Dawkins (1976), p. 3.

7. Frans de Waal (2013).

8. Mary Midgley (1995) and (2010); Gregory McElwain (2020).

9. Frans de Waal (2006).

10. Inbal Ben-Ami Bartal et al. (2011).

11. Melanie Killen and Elliot Turiel (1991); Cary Roseth (2018).

12. Rutger Bregman (2019).

13. Toni Morrisson (2019).

14. Henry Nicholls (2014).

15. Hans Kummer (1995), p. xviii.

16. Hans Kummer (1995), p. 193; Christian Bachmann and Hans Kummer (1980).

17. Jared Diamond (1992).

18. K.R.L. Hall and Irven DeVore (1965).

19. Thelma Rowell (1974), p. 44.

20. Curt Busse (1980).

21. Vinciane Despret (2009).

22. Barbara Smuts (1985).

23. Robert Seyfarth and Dorothy Cheney (2012); Lydia Denworth (2019).

24. Nga Nguyen et al. (2009).

25. Donna Haraway (1989), pp. 150, 154.

26. Matt Cartmill (1991).

27. Jeanne Altmann (1974).

28. Alison Jolly (1999), p. 146.

29. Linda Marie Fedigan (1994).

30. Shirley Strum (2012).

CHAPTER 5. BONOBO SISTERHOOD

1. Lola ya Bonobo website: www.bonobos.org.

2. Nahoko Tokuyama et al. (2019).

3. Claudine André (2006), pp.167–74.

4. For Mimi's introduction, see L'ange des bonobos, August 13, 2019, youtube
   .com/watch?v=VedUkzx7YOk.

5. Eva Maria Luef et al. (2016).

6. Robert Yerkes (1925), p. 244.

7. Adrienne Zihlman et al. (1978).

8. Jacques Vauclair and Kim Bard (1983).

9. Stephen Jay Gould (1977); Robert Bednarik (2011).

10. Frans de Waal (1989).

11. Elisabetta Palagi and Elisa Demuru (2017).

12. Sven Grawunder et al. (2018).

13. Eduard Tratz and Heinz Heck (1954), p. 99 (translated from German).

14. Kay Prüfer et al. (2012).

15. Nick Patterson et al. (2006); but see Masato Yamamichi et al. (2012).

16. Harold Coolidge (1933), p. 56; Rui Diogo et al. (2017).

17. Takayoshi Kano (1992); Frans de Waal (1987).

18. Zanna Clay and Frans de Waal (2013).

19. Robert Ardrey (1961).

20. Matt Cartmill (1993).

21. Gen'ichi Idani (1990); Takayoshi Kano (1992).

22. Steven Pinker (2011), p. 39; Richard Wrangham (2019), p. 98.

23. Adam Rutherford (2018), p. 105; Craig Stanford (1998).

24. Frans de Waal (1997), with Frans Lanting's photography.

25. Amy Parish (1993).

26. Takayoshi Kano (1998), p. 410.

27. Takeshi Furuichi (2019).

28. Martin Surbeck and Gottfried Hohmann (2013).

29. Takeshi Furuichi et al. (2014).

30. Frans de Waal (2016).

31. Natalie Angier (1997).

32. Martin Surbeck et al. (2017).

33. Gottfried Hohmann and Barbara Fruth (2011); Nahoko Tokuyama and Takeshi Furuichi (2017); Tokuyama et al. (2019).

34. Takeshi Furuichi (2019), p. 62.

35. Benjamin Beck (2019).

36. Sydney Richards, Primate heroes: PASA's amazing women leaders, Pan African Sanctuary Alliance, n.d., pasa.org/awareness/primate-heroes-pasas -amazing-women-leaders.

CHAPTER 6. SEXUAL SIGNALS

1. Desmond Morris (1967), p. 5.

2. Detlev Ploog and Paul MacLean (1963).

3. Wolfgang Wickler (1969); Desmond Morris (1977).
4. Tanya Vacharkulksemsuka et al. (2016).
5. For more about female choice, see Chapter 7.
6. Edgar Berman (1982).
7. Richard Harlan (1827); Anna Maerker (2005).
8. Emmanuele Jannini et al. (2014); Rachel Pauls (2015); Nicole Prause et al. (2016).
9. Thomas Laqueur (1990), p. 236.
10. Natalie Angier (2000).
11. Elisabeth Lloyd (2005); The ideas interview: Elisabeth Lloyd, *Guardian,* September 26, 2005, www.theguardian.com/science/2005/sep/26/genderissues.technology.
12. Steven Jay Gould (1993).
13. Helen O'Connell et al. (2005); Vincenzo Puppo (2013).
14. Dara Orbach and Patricia Brennan (2021).
15. David Goldfoot et al. (1980).
16. Sue Savage-Rumbaugh and Beverly Wilkerson (1978); Frans de Waal (1987).
17. Anne Pusey (1980); Elisa Demuru et al. (2020).
18. Frans de Waal and Jennifer Pokorny (2008).
19. Willemijn van Woerkom and Mariska Kret (2015); Mariska Kret and Masaki Tomonaga (2016).
20. Richard Prum (2017).
21. Elizabeth Cashdan (1998); Rebecca Nash et al. (2006).
22. Karl Grammer et al. (2005); Martie Haselton et al. (2007).
23. Wolfgang Köhler (1925), p. 84.
24. Robert Yerkes (1925), p. 67.
25. Edwin van Leeuwen et al. (2014).
26. Warren Roberts and Mark Krause (2002).
27. My translation of Jürgen Lethmate and Gerti Dücker (1973), p. 254.
28. Vernon Reynolds (1967).
29. William McGrew and Linda Marchant (1998).
30. Robert Yerkes (1941).
31. Ruth Herschberger (1948), p. 10.
32. Jane Goodall (1986), p. 483.
33. Kimberly Hockings et al. (2007).
34. Vicky Bruce and Andrew Young (1998); Alessandro Cellerino et al. (2004); Richard Russell (2009).

CHAPTER 7. THE MATING GAME

1. Abraham Maslow (1936); Dallas Cullen (1997).
2. Frans de Waal and Lesleigh Luttrell (1985).
3. Martin Curie-Cohen et al. (1983); Bonnie Stern and David Glenn Smith (1984); John Berard et al. (1994); Susan Alberts et al. (2006).
4. Simon Townsend et al. (2008).
5. St. George Mivart (1871), in Richard Prum (2015).
6. Claude Lévi-Strauss (1949).
7. Olin Bray et al. (1975).
8. Tim Birkhead and John Biggins (1987); Bridget Stutchbury et al. (1997); David Westneat and Ian Stewart (2003); Kathi Borgmann (2019).
9. Nicholas Davies (1992); Steve Connor (1995).
10. Steven Verseput, New Kim, de duif die voor 1,6 miljoen euro naar China ging, NRC, November 20, 2020 (Dutch).
11. Patricia Gowaty (1997).
12. In biology, immediate motives are called "proximate" causation of behavior, and evolutionary reasons "ultimate" causation; Ernst Mayr (1982).
13. Gregor Mendel's results, first published in 1865, were rediscovered in 1900.
14. Malcolm Potts and Roger Short (1999), p. 319.
15. Heather Rupp and Kim Wallen (2008); Ruben Arslan et al. (2018).
16. Caroline Tutin (1979); Kees Nieuwenhuijsen (1985).
17. Janet Hyde and John DeLamater (1997).
18. Roy Baumeister et al. (2001).
19. Sheila Murphy (1992); Roy Baumeister (2010).
20. Tom Smith (1991); Michael Wiederman (1997).
21. Michele Alexander and Terri Fisher (2003).
22. Angus Bateman (1948); Robert Trivers (1972).
23. E. O. Wilson (1978), p. 125.
24. Patricia Gowaty et al. (2012); Thierry Hoquet et al. (2020).
25. Monica Carosi and Elisabetta Visalberghi (2002).
26. Susan Perry (2008), p. 166.
27. Sarah Blaffer Hrdy (1977).
28. Yukimaru Sugiyama (1967).
29. Frans de Waal (1982); Jane Goodall (1986).
30. Sarah Blaffer Hrdy (2000).
31. Carson Murray et al. (2007).
32. Takayoshi Kano (1992), p. 208.

33. Frans de Waal (1997); Amy Parish and Frans de Waal (2000).
34. Martin Daly and Margo Wilson (1988).
35. Stephen Beckerman et al. (1998).
36. Meredith Small (1989); Sarah Blaffer Hrdy (1999), p. 251.
37. Aimee Ortiz (2020).

CHAPTER 8. VIOLENCE

1. Patricia Tjaden and Nancy Thoennes (2000).
2. David Watts et al. (2006).
3. Toshisada Nishida (1996) and (2012).
4. Jane Goodall (1979); Richard Wrangham and Dale Peterson (1996); Warren Manger, Jane Goodall: I thought chimps were like us only nicer, but we inherited our dark evil side from them, *Mirror* (UK), March 12, 2018, www.mirror.co.uk/news/world-news/jane-goodall-chimpanzees-evil -apes-12170154.
5. Michael Wilson et al. (2014).
6. Global data for 2012 in "Homicide and Gender," 2015 report by UN Office on Drugs and Crime, https://heuni.fi/documents/47074104/49490570/ Homicide_and_Gender.pdf.
7. Pink Floyd's 1987 album, *A Momentary Lapse of Reason.*
8. Joshua Goldstein (2001); Adam Jones (2002).
9. Oriel FeldmanHall et al. (2016).
10. Hannah Arendt (1984); Daniel Goldhagen (1996); Jonathan Harrison (2011); Nestar Russell (2019).
11. Elizabeth Brainerd (2016).
12. Barbara Smuts (2001), p. 298.
13. Eugene Linden (2002).
14. Martin Muller et al. (2009) and (2011); Joseph Feldblum et al. (2014).
15. Rape addendum, FBI's Uniform Crime Reporting (2013), https://ucr.fbi .gov/crime-in-the-u.s/2013/crime-in-the-u.s.-2013/rape-addendum/rape_ addendum_final.
16. Jane Goodall (1986).
17. Shiho Fujita and Eiji Inoue (2015), p. 487.
18. Julie Constable et al. (2001).
19. John Mitani and Toshisada Nishida (1993).
20. Christophe Boesch (2009).

21. Christophe Boesch and Hedwige Boesch-Achermann (2000); Rebecca Stumpf and Christophe Boesch (2010).

22. Patricia Tjaden and Nancy Thoennes (2000).

23. Brad Boserup et al. (2020).

24. Biruté Galdikas (1995).

25. Carel van Schaik (2004), p. 76.

26. Cheryl Knott and Sonya Kahlenberg (2007).

27. Jack Weatherford (2004), p. 111.

28. Heidi Stöckl et al. (2013).

29. Preventing sexual violence, Centers for Disease Control and Prevention, n.d., www.cdc.gov/violenceprevention/sexualviolence/fastfact.html.

30. Susan Brownmiller (1975), p. 14.

31. Randy Thornhill and Craig Palmer (2000).

32. Patricia Tjaden and Nancy Thoennes (2000).

33. Cheryl Brown Travis (2003); Joan Roughgarden (2004).

34. Frans de Waal (2000).

35. Eric Smith et al. (2001).

36. Gert Stulp et al. (2013); George Yancey and Michael Emerson (2016).

37. Aaron Sell et al. (2017).

38. Gayle Brewer and Sharon Howarth (2012); Robert Deaner et al. (2015).

39. Siobhan Heanue, Indian women form a gang and roam their village, punishing men for their bad behaviour, ABC News, August 3, 2019, www.abc .net.au/news/2019-08-04/indian-women-get-together-to-punish-men -who-wrong-them/11369326.

40. Barbara Smuts (1992); Barbara Smuts and Robert Smuts (1993).

41. Marianne Schnall, Interview with Gloria Steinem on equality, her new memoir, and more, Feminist.com, c. 2016, www.feminist.com/resources/ artspeech/interviews/gloriasteineminterview.

CHAPTER 9. ALPHA (FE)MALES

1. Rudolf Schenkel (1947).

2. Elspeth Reeve (2013).

3. On Solly Zuckerman, see Chapter 4; Robert Ardrey (1961), p. 144.

4. Quincy Wright (1965), p. 100.

5. Samuel Bowles and Herbert Gintis (2003); Michael Morgan and David Carrier (2013).

6. Napoleon Chagnon (1968); Richard Wrangham and Dale Peterson (1996).
7. Doug Fry (2013).
8. Mark Foster et al. (2009).
9. For the rhesus monkeys Spickles and Orange, see Chapter 7.
10. Kinji Imanishi (1960), quoted in Linda Fedigan (1982), p. 91.
11. Christina Cloutier Barbour, unpublished data.
12. Steffen Foerster et al. (2016).
13. Frans de Waal (1986).
14. Toshisada Nishida and Kazuhiko Hosaka (1996).
15. Joseph Henrich and Francisco Gil-White (2001).
16. Victoria Horner et al. (2010).
17. Sean Wayne (2021).
18. Jane Goodall (1990).
19. Teresa Romero et al. (2010).
20. Robert Sapolsky (1994).
21. David Watts et al. (2000).
22. Christopher Boehm (1999), p. 27.
23. Frans de Waal (1984); Christopher Boehm (1994); Claudia von Rohr et al. (2012).
24. Jessica Flack et al. (2005).
25. Rob Slotow et al. (2000); Caitlin O'Connell (2015).
26. Aaron Sandel et al. (2020).
27. Nancy Vaden-Kierman et al. (1995); Stephen Demuth and Susan Brown (2004); Sarah Hill et al. (2016); The proof is in: Father absence harms children, National Fatherhood Initiative, n.d., www.fatherhood.org/father-absence-statistic.
28. Martha Kirkpatrick (1987).
29. Terry Maple (1980); S. Utami Atmoko (2000); Anne Maggioncalda et al. (2002); Carel van Schaik (2004).
30. Sarah Romans et al. (2003); Bruce Ellis et al. (2003); Anthony Bogaert (2005); James Chisholm et al. (2005); Julianna Deardorff et al. (2010).
31. Christophe Boesch (2009).
32. Takeshi Furuichi (1997).
33. Martin Surbeck et al. (2019); Ed Yong (2019).
34. Leslie Peirce (1993).
35. Stewart McCann (2001); Nancy Blaker et al. (2013).
36. Nicholas Kristof, What the pandemic reveals about the male ego, *New York Times,* June 13, 2020.

37. Viktor Reinhardt et al. (1986).
38. Marianne Schmid Mast (2002) and (2004).
39. Christopher Boehm (1993) and (1999); Harold Leavitt (2003).
40. Barbara Smuts (1987); Rebecca Lewis (2018).

CHAPTER 10. KEEPING THE PEACE

1. Alessandro Cellerino et al. (2004).
2. Cal State Northridge professor charged with peeing on colleague's door, Associated Press, January 27, 2011, https://www.scpr.org/news/2011/01 /27/23415/cal-state-northridge-professor-charged-peeing-coll/.
3. Elizabeth Cashdan (1998).
4. Idan Frumin et al. (2015).
5. Shelley Taylor (2002); Lydia Denworth (2020), p.157.
6. Amanda Rose and Karen Rudolph (2006).
7. Jeffrey Hall (2011); Lydia Denworth (2020).
8. Marilyn French (1985), p. 271.
9. Phyllis Chesler (2002).
10. Matthew Gutmann (1997), p. 385; Samuel Bowles (2009).
11. Lionel Tiger (1969), p. 259.
12. Daniel Balliet et al. (2011).
13. *Steve Martin and Martin Short: An Evening You Will Forget for the Rest of Your Life* (Netflix, 2018).
14. Gregory Silber (1986); Caitlin O'Connell (2015).
15. Peter Marshall et al. (2020).
16. Joshua Goldstein (2001); Dieter Leyk et al. (2007).
17. Alexandra Rosati et al. (2020).
18. Sarah Blaffer Hrdy (1981), p. 129.
19. Anne Campbell (2004).
20. Kirsti Lagerspetz et al. (1988).
21. Rachel Simmons (2002); Emily White (2002); Rosalind Wiseman (2016).
22. Margaret Atwood (1989), p. 166.
23. Kai Björkqvist et al. (1992).
24. Janet Lever (1976); Zick Rubin (1980); Joyce Benenson and Athena Christakos (2003).
25. Joyce Benenson and Richard Wrangham (2016); Joyce Benenson et al. (2018).

26. Frans de Waal and Angeline van Roosmalen (1979).
27. Filippo Aureli and Frans de Waal (2000); Frans de Waal (2000); Kate Arnold and Andrew Whiten (2001); Roman Wittig and Christophe Boesch (2005).
28. Frans de Waal (1993); Sonja Koski et al. (2007).
29. Orlaith Fraser and Filippo Aureli (2008).
30. Filippo Aureli and Frans de Waal (2000).
31. Elisabetta Palagi et al. (2004); Zanna Clay and Frans de Waal (2015).
32. Susan Nolen-Hoeksema et al. (2008).
33. Neil Brewer et al. (2002); Julia Bear et al. (2014).
34. Sarah Blaffer Hrdy (2009).
35. Sandra Boodman (2013).
36. Laura Jones et al. (2018).
37. Ingo Titze and Daniel Martin (1998).
38. Monica Hesse (2019); David Moye (2019).
39. Charlotte Riley (2019).
40. Deirdre McCloskey (1999); Tara Bahrampour (2018); Charlotte Alter (2020).
41. Thomas Page McBee (2016).
42. Sarah Collins (2000); David Andrew Puts et al. (2007); Casey Klofstad et al. (2012).
43. Alecia Carter et al. (2018).

CHAPTER 11. NURTURANCE

1. Patricia Churchland (2019), p. 22.
2. Trevor Case et al. (2006); Johan Lundström et al. (2013).
3. Inna Schneiderman et al. (2012); Sara Algoe et al. (2017).
4. Christopher Krupenye et al. (2016).
5. Frans de Waal (1996a); Shinya Yamamoto et al. (2009).
6. Stephanie Musgrave et al. (2016).
7. Christophe Boesch and Hedwige Boesch-Achermann (2000); Frans de Waal (2009).
8. Frans de Waal (2008).
9. James Burkett et al. (2016); Frans de Waal and Stephanie Preston (2017).
10. Frans de Waal (1996b).
11. William Hopkins (2004); Brenda Todd and Robin Banerjee (2018); Gillian Forrester et al. (2019).
12. William Hopkins and Mieke de Lathouwers (2006).

13. Anthony Volk (2009).
14. Judith Blakemore (1990) and (1998); Dario Maestripieri and Suzanne Pelka (2002).
15. Lev Vygotsky (1935) quoted in Anna Chernaya (2014), p. 186.
16. Chapter 1 and Sonya Kahlenberg and Richard Wrangham (2010).
17. Melvin Konner (1976); Carolyn Edwards (1993), p. 331, and (2005).
18. Jane Lancaster (1971), p. 170.
19. Lynn Fairbanks (1990) and (1993); Joan Silk (1999); Rebecca Hermann et al. (2003); Ulia Bădescu et al. (2015).
20. Herman Dienske et al. (1980).
21. Alison Flemming et al. (2002); Ioana Carcea et al. (2020).
22. Charles Darwin, Notebook D (1838), https://tinyurl.com/2xbmfjsd, p. 154; Joseph Lonstein and Geert de Vries (2000).
23. Charles Snowdon and Toni Ziegler (2007).
24. Susan Lappan (2008).
25. Kimberley Hockings et al. (2006).
26. Jill Pruetz (2011).
27. Christophe Boesch et al. (2010).
28. Rachna Reddy and John Mitani (2019).
29. Gen'ichi Idani (1993).
30. For partible paternity, see Chapter 7.
31. Bhismadev Chakrabarti and Simon Baron-Cohen (2006), p. 408; Linda Rueckert et al. (2011); Frans de Waal and Stephanie Preston (2017).
32. Carolyn Zahn-Waxler et al. (1992).
33. Marie Lindegaard et al. (2017).
34. Martin Schulte-Rüther et al. (2008); Birgit Derntl et al. (2010).
35. Shir Atzil et al. (2012); Ruth Feldman et al. (2019).
36. Sarah Schoppe-Sullivan et al. (2021).
37. Carol Clark, Five surprising facts about fathers, Emory University, http://news.emory.edu/features/2019/06/five-facts-fathers.
38. James Rilling and Jennifer Mascaro (2017).
39. Margaret Mead (1949), p. 145.
40. Sarah Blaffer Hrdy (2009), p. 109.
41. Frans de Waal (2013), p. 139; Elisa Demuru et al. (2018).
42. Lynn Fairbanks (2000).
43. Darren Croft et al. (2017).
44. Kristen Hawkes and James Coxworth (2013); Simon Chapman et al. (2019).

45. Charles Weisbard and Robert Goy (1976).
46. Zoë Goldsborough et al. (2020).
47. Christophe Boesch (2009), p. 48.

CHAPTER 12. SAME-SEX SEX

1. Maggie Hiufu Wong, Incest and affairs of Japan's scandalous penguins, CNN, December 5, 2019, www.cnn.com/travel/article/aquarium
   -penguins-japan.
2. Douglas Russell et al. (2012).
3. Pinguin-Damen sollen schwule Artgenossen bezirzen, *Kölner Stadt-
   Anzeiger,* August 1, 2005 (German).
4. *APA Dictionary of Psychology,* 2nd ed. (Washington, DC: American Psychological Association, 2015).
5. *Lawrence v. Texas,* 539 U.S. 558, 2003; Dick Swaab (2010).
6. Jonathan Miller, New love breaks up a 6-year relationship at the zoo, *New
   York Times,* September 24, 2005.
7. Gwénaëlle Pincemy et al. (2010), p. 1211.
8. Quinn Gawronski, Gay penguins at London aquarium are raising "genderless" chick, September 10, 2019, https://tinyurl.com/car3ce8x.
9. Paul Vasey (1995).
10. Jean-Baptiste Leca et al. (2014).
11. Jake Brooker et al. (2021).
12. Frank Beach (1949).
13. Clellan Ford and Frank Beach (1951); Neel Burton (2015).
14. Bruce Bagemihl (1999); Alan Dixon (2010).
15. Linda Wolfe (1979); Gail Vines (1999).
16. Bruce Bagemihl (1999), p. 117.
17. Frans de Waal (1987) (1997).
18. Takayoshi Kano (1992).
19. Liza Moscovice et al. (2019); Elisabetta Palagi et al. (2020).
20. Zanna Clay and Frans de Waal (2015).
21. Dick Swaab and Michel Hofman (1990); Dick Swaab (2010).
22. E. O. Wilson (1978), p. 167.
23. Simon LeVay (1991); Janet Halley (1994); Elizabeth Wilson (2000).
24. William Byne et al. (2001).
25. Ivanka Savic and Per Lindström (2008); Andy Coghlan (2008).
26. Ivanka Savic et al. (2005); Wen Zhou et al. (2014).

27. Bruce Bagemihl (1999); Charles Roselli et al. (2004).
28. Niklas Långström et al. (2010); Andrea Ganna et al. (2019).
29. Ritch Savin-Williams and Zhana Vrangalova (2013); Jeremy Jabbour et al. (2020).
30. Alfred Kinsey et al. (1948), p. 639.
31. Milton Diamond, Nature loves variety, society hates it, interview, December 24, 2013, www.youtube.com/watch?v=6MvNisJ7FoQ.
32. Adam Rutherford (2020).
33. Simon LeVay (1996), p. 209.
34. David Greenberg (1988); Pieter Adriaens and Andreas de Block (2006).
35. Malcolm Potts and Roger Short (1999), p. 74.
36. Sergey Gavrilets and William Rice (2006).
37. Benedict Regan et al. (2001).
38. Frans de Waal (2009); Cammie Finch (2016).
39. Cindy Meston and David Buss (2007).
40. Joan Roughgarden (2017), p. 512.

CHAPTER 13. THE TROUBLE WITH DUALISM

1. Mary Midgley (1995).
2. Robert Sapolsky (1997); Rebecca Jordan-Young and Katrina Karkazis (2019).
3. Gina Rippon (2019).
4. Simon Baron-Cohen in The gendered brain debate (podcast), How To Academy, n.d., howtoacademy.com/podcasts/the-gendered-brain-debate.
5. Margaret McCarthy (2016); Erin Hecht et al. (2020).
6. Frans de Waal (2001); Victoria Horner and Frans de Waal (2009).
7. The Gospel of Thomas, Sacred-Texts.com, www.sacred-texts.com/chr/thomas.htm.
8. Antonio Damasio (1999), p. 143.
9. Brian Calvert (1975); Elizabeth Spelman (1982).
10. Mark O'Connell (2017).
11. Elizabeth Spelman (1982), p. 120.
12. Elizabeth Wilson (1998).

BIBLIOGRAPHY

Adriaens, P. R., and A. de Block. 2006. The evolution of a social construction: The case of male homosexuality. *Perspectives in Biology and Medicine* 49:570–85.

Alberts, S. C., J. C. Buchan, and J. Altmann. 2006. Sexual selection in wild baboons: From mating opportunities to paternity success. *Animal Behaviour* 72:1177–96.

Alexander, G. M., and M. Hines. 2002. Sex differences in response to children's toys in nonhuman primates. *Evolution and Human Behavior* 23:467–79.

Algoe, S. B., L. E. Kurtz, and K. Grewen. 2017. Oxytocin and social bonds: The role of oxytocin in perceptions of romantic partners' bonding behavior. *Psychological Science* 28:1763–72.

Alsop, R., A. Fitzsimons, and K. Lennon. 2002. The social construction of gender. In *Theorizing Gender,* ed. R. Alsop, A. Fitzsimons and K. Lennon, pp. 64–93. Malden, MA: Blackwell.

Alter, C. 2020. "Cultural sexism in the world is very real when you've lived on both sides of the coin." *Time,* n.d., time.com/transgender-men-sexism

Altmann, J. 1974. Observational study of behavior. *Behaviour* 49:227–65.

André, C. 2006. *Une Tendresse Sauvage.* Paris: Calmann-Lévy (French).

Angier, N. 1997. Bonobo society: Amicable, amorous and run by females. *New York Times,* April 11, 1997, p. C4.

———. 2000. *Woman: An Intimate Geography.* New York: Anchor Books.

Ardrey, R. 2014 (orig. 1961). *African Genesis: A Personal Investigation into the Animal Origins and Nature of Man.* N.p.: StoryDesign.

Arendt, H. 1984. *Eichmann in Jerusalem: A Report on the Banality of Evil.* New York: Penguin.

Arnold, K., and A. Whiten. 2001. Post-conflict behaviour of wild chimpanzee in the Budongo Forest, Uganda. *Behaviour* 138:649–90.

Arslan, R. C., et al. 2018. Using 26,000 diary entries to show ovulatory changes in sexual desire and behavior. *Journal of Personality and Social Psychology.* Advance online publication.

Atwood, M. E. 1989. *Cat's Eye.* New York: Doubleday.

Atzil, S., et al. 2012. Synchrony and specificity in the maternal and the paternal brain: Relations to oxytocin and vasopressin. *Journal of the American Academy of Child and Adolescent Psychiatry* 51:798–811.

Aureli, F., and F. B. M. de Waal. 2000. *Natural Conflict Resolution.* Berkeley: University of California Press.

Bachmann, C., and H. Kummer. 1980. Male assessment of female choice in hamadryas baboons. *Behavioral Ecology and Sociobiology* 6:315–21.

Bădescu, J., et al. 2015. Female parity, maternal kinship, infant age and sex influence natal attraction and infant handling in a wild colobine. *American Journal of Primatology* 77:376–87.

Bagemihl, B. 1999. *Biological Exuberance: Animal Homosexuality and Natural Diversity.* New York: St. Martin's.

Bahrampour, T. 2018. Crossing the divide. *Washington Post,* July 20, 2018.

Balliet, D., et al. 2011. Sex differences in cooperation: A meta-analytic review of social dilemmas. *Psychological Bulletin* 137:881–909.

Bao, A.-M., and D. F. Swaab. 2011. Sexual differentiation of the human brain: Relation to gender identity, sexual orientation and neuropsychiatric disorders. *Frontiers in Neuroendocrinology* 32:214–26.

Barrett, L. F., L. Robin, and P. R. Pietromonaco. 1998. Are women the more emotional sex? Evidence from emotional experiences in social context. *Cognition and Emotion* 12:555–78.

Bartal, I. B-A., J. Decety, and P. Mason. 2011. Empathy and pro-social behavior in rats. *Science* 334:1427–30.

Bateman, A. J. 1948. Intra-sexual selection in Drosophila. *Heredity* 2:349–68.

Baumeister, R. F. 2010. The reality of the male sex drive. *Psychology Today,* December 10, 2018.

Baumeister, R. F., K. R. Catanese, and K. D. Vohs. 2001. Is there a gender difference in strength of sex drive? Theoretical views, conceptual distinctions, and a review of relevant evidence. *Personality and Social Psychology Review* 5:242–73.

Baumeister, R. F., K. D. Vohs, and D. C. Funder. 2007. Psychology as the science of self-reports and finger movements: Whatever happened to actual behavior? *Perspectives on Psychological Science* 2:396–403.

Beach, F. A. 1949. A cross-species survey of mammalian sexual behavior. In *Psychosexual Development in Health and Disease,* ed. P. H. Hoch and J. Zubin, pp. 52–78. New York: Grune and Stratton.

Bear, J. B., L. R. Weingart, and G. Todorova. 2014. Gender and the emotional experience of relationship conflict: The differential effectiveness of avoidant conflict management. *Negotiation and Conflict Management Research* 7:213–31.

Beck, B. B. 2019. *Unwitting Travelers: A History of Primate Reintroduction.* Berlin, MD: Salt Water Media.

Beckerman, S., et al. 1998. The Barí Partible Paternity Project: Preliminary results. *Current Anthropology* 39:164–68.

Bednarik, R. G. 2011. *The Human Condition.* New York: Springer.

Benenson, J. F., and A. Christakos. 2003. The greater fragility of females' versus males' closest same-sex friendships. *Child Development* 74:1123–29.

Benenson, J. F., and R. W. Wrangham. 2016. Differences in post-conflict affiliation following sports matches. *Current Biology* 26:2208–12.

Benenson, J. F., et al. 2018. Competition elicits more physical affiliation between male than female friends. *Scientific Reports* 8:8380.

Berard, J. D., P. Nurnberg, J. T. Epplen, and J. Schmidtke. 1994. Alternative reproductive tactics and reproductive success in male rhesus macaques. *Behaviour* 129:177–200.

Berman, E. 1982. *The Compleat Chauvinist: A Survival Guide for the Bedeviled Male.* New York: Macmillan.

Biba, E. 2019. In real life, Simba's mom would be running the pride. *National Geographic,* July 8, 2019.

Birkhead, T. R., and J. D. Biggins. 1987. Reproductive synchrony and extra-pair copulation in birds. *Ethology* 74:320–34.

Björkqvist, K., et al. 1992. Do girls manipulate and boys fight? Developmental trends in regard to direct and indirect aggression. *Aggressive Behavior* 18:117–27.

Black, J. M. 1996. *Partnerships in Birds: The Study of Monogamy.* Oxford: Oxford University Press.

Blakemore, J. E. O. 1990. Children's nurturant interactions with their infant siblings: An exploration of gender differences and maternal socialization. *Sex Roles* 22:43–57.

———. 1998. The influence of gender and parental attitudes on preschool children's interest in babies: Observations in natural settings. *Sex Roles* 38:73–94.

Blaker, N. M., et al. 2013. The height leadership advantage in men and women: Testing evolutionary psychology predictions about the perceptions of tall leaders. *Group Processes and Intergroup Relations* 16:17–27.

Boehm, C. 1993. Egalitarian behavior and reverse dominance hierarchy. *Current Anthropology* 34:227–54.

———. 1994. Pacifying interventions at Arnhem Zoo and Gombe. In *Chimpanzee Cultures,* ed. R. W. Wrangham et al., pp. 211–26. Cambridge, MA: Harvard University Press.

———. 1999. *Hierarchy in the Forest: The Evolution of Egalitarian Behavior.* Cambridge, MA: Harvard University Press.

Boesch, C. 2009. *The Real Chimpanzee: Sex Strategies in the Forest.* Cambridge, UK: Cambridge University Press.

Boesch, C., and H. Boesch-Achermann. 2000. *The Chimpanzees of the Taï Forest: Behavioural Ecology and Evolution.* Oxford: Oxford University Press.

Boesch, C., et al. 2010. Altruism in forest chimpanzees: The case of adoption. *PLoS ONE* 5:e8901.

Bogaert, A. F. 2005. Age at puberty and father absence in a national probability sample. *Journal of Adolescence* 28:541–46.

Bono, A. E. J., et al. 2018. Payoff- and sex-biased social learning interact in a wild primate population. *Current Biology* 28:2800–5.

Boodman, S. G. 2013. Anger management courses are a new tool for dealing with out-of-control doctors. *Washington Post,* March 4, 2013.

Borgmann, K. 2019. The forgotten female: How a generation of women scientists changed our view of evolution. *All About Birds,* June 17, 2019.

Boserup, B., et al. 2020. Alarming trends in US domestic violence during the COVID-19 pandemic. *American Journal of Emergency Medicine* 38:2753–55.

Bouazzouni, N. 2017. *Faiminisme: Quand le sexisme passe à table.* Paris: Nouriturfu (French).

Bowles, S. 2009. Did warfare among ancestral hunter-gatherers affect the evolution of human social behaviors? *Science* 324:1293–98.

Bowles, S., and H. Gintis. 2003. The origins of human cooperation. In *The Genetic and Cultural Origins of Cooperation*, ed. P. Hammerstein, pp. 429–44. Cambridge, MA: MIT Press.

Brainerd, E. 2016. *The Lasting Effect of Sex Ratio Imbalance on Marriage and Family: Evidence from World War II in Russia.* IZA Discussion Paper no. 10130.

Bray, O. E., J. J. Kennelly, and J. L. Guarino. 1975. Fertility of eggs produced on territories of vasectomized red-winged blackbirds. *Wilson Bulletin* 87:187–95.

Bregman, R. 2019. *De Meeste Mensen Deugen: Een Nieuwe Geschiedenis van de Mens.* Amsterdam: De Correspondent (Dutch).

Brewer, G., and S. Howarth. 2012. Sport, attractiveness, and aggression. *Personality and Individual Differences* 53:640–43.

Brewer, N., P. Mitchell, and N. Weber. 2002. Gender role, organizational status, and conflict management styles. *International Journal of Conflict Management* 13:78–94.

Brooker, J. S., C. E. Webb, and Z. Clay. 2021. Fellatio among male sanctuary-living chimpanzees during a period of social tension. *Behaviour* 158:77–87.

Brownmiller, S. 1975. *Against Our Will: Men, Women and Rape.* New York: Simon and Schuster.

Bruce, V., and A. Young 1998. *In the Eye of the Beholder: The Science of Face Perception.* Oxford: Oxford University Press.

Burkett, J. P., et al. 2016. Oxytocin-dependent consolation behavior in rodents. *Science* 351:375–78.

Burton, N. 2015. When homosexuality stopped being a mental disorder. *Psychology Today,* September 18, 2015.

Busse, C. 1980. Leopard and lion predation upon chacma baboons living in the Moremi Wildlife Reserve. *Botswana Notes and Records* 12:15–21.

Butler, J. 1986. Sex and gender in Simone de Beauvoir's *Second Sex. Yale French Studies* 72:35–49.

———. 1988. Performative acts and gender constitution: An essay in phenomenology and feminist theory. *Theatre Journal* 40:519–31.

Byne, W., et al. 2001. The interstitial nuclei of the human anterior hypothalamus: An investigation of variation within sex, sexual orientation and HIV status. *Hormones and Behavior* 40:86–92.

Calvert, B. 1975. Plato and the equality of women. *Phoenix* 29:231–43.

Campbell, A. 2004. Female competition: Causes, constraints, content, and contexts. *Journal of Sex Research* 41:16–26.

Carcea, I., et al. 2020. Oxytocin neurons enable social transmission of maternal behavior. *BioRxiv,* www.biorxiv.org/content/10.1101/845495v1.

Carlin, J. 1995. How Newt aped his way to the top. *Independent,* May 30, 1995.

Carosi, M., and E. Visalberghi. 2002. Analysis of tufted capuchin courtship and sexual behavior repertoire: Changes throughout the female cycle and female interindividual differences. *American Journal of Physical Anthropology* 118:11–24.

Carson, R. 1962. *Silent Spring.* New York: Houghton Mifflin.

Carter, A. J., et al. 2018. Women's visibility in academic seminars: Women ask fewer questions than men. *PLoS ONE* 13:e0202743.

Cartmill, M. 1991. Review of *Primate Visions,* by Donna Haraway. *International Journal of Primatology* 12:67–75.

———. 1993. *A View to a Death in the Morning.* Cambridge, MA: Harvard University Press.

Case, T. I., B. M. Repacholi, and R. J. Stevenson. 2006. My baby doesn't smell as bad as yours: The plasticity of disgust. *Evolution and Human Behavior* 27:357–65.

Cashdan, E. 1998. Are men more competitive than women? *British Journal of Social Psychology* 37:213–29.

Cellerino, A., D. Borghetti, and F. Sartucci. 2004. Sex differences in face gender recognition in humans. *Brain Research Bulletin* 63:443–49.

Chagnon, N. A. 1968. *Yanomamö: The Fierce People.* New York: Holt, Rinehart and Winston.

Chakrabarti, B., and S. Baron-Cohen. 2006. Empathizing: Neurocognitive developmental mechanisms and individual differences. *Progress in Brain Research* 156:403–17.

Chapman, S. N., et al. 2019. Limits to fitness benefits of prolonged post-reproductive lifespan in women. *Current Biology* 29:645–50.

Chernaya, A. 2014. Girls' plays with dolls and doll-houses in various cultures. In *Proceedings from the 21st Congress of the International Association for Cross-Cultural Psychology,* ed. L. T. B. Jackson et al.

Chesler, P. 2002. *Woman's Inhumanity to Woman.* New York: Nation Books.

Chisholm, J. S., et al. 2005. Early stress predicts age at menarche and first birth, adult attachment, and expected lifespan. *Human Nature* 16:233–65.

Christie, A. 1933. *The Hound of Death and Other Stories.* London: Odhams Press.

Christopher, B. 2016. The massacre at Monkey Hill. *Priceonomics,* n.d., priceonomics.com/the-massacre-at-monkey-hill.

Churchland, P. S. 2019. *Conscience: The Origins of Moral Intuition.* New York: Norton.

Clay, Z., and F. B. M. de Waal. 2013. Development of socio-emotional competence in bonobos. *Proceedings of the National Academy of Sciences USA* 110:18121–26.

———. 2015. Sex and strife: Post-conflict sexual contacts in bonobos. *Behaviour* 152:313–34.

Coghlan, A. 2008. Gay brains structured like those of the opposite sex. *New Scientist,* June 16, 2008.

Colapinto, J. 2000. *As Nature Made Him: The Boy Who Was Raised as a Girl.* New York: Harper.

Collins, S. A. 2000. Men's voices and women's choices. *Animal Behaviour* 60:773–80.

Connellan, J., et al. 2000. Sex differences in human neonatal social perception. *Infant Behavior and Development* 23:113–18.

Connor, S. 1995. Reflection: Why bishops are like apes. *Independent,* May 18, 1995.

Constable, J. L., et al. 2001. Noninvasive paternity assignment in Gombe chimpanzees. *Molecular Ecology* 10:1279–300.

Coolidge, H. J. 1933. *Pan paniscus:* Pygmy chimpanzee from south of the Congo River. *American Journal of Physical Anthropology* 18:1–57.

Croft, D. P., et al. 2017. Reproductive conflict and the evolution of menopause in killer whales. *Current Biology* 27:298–304.

Cullen, D. 1997. Maslow, monkeys, and motivation theory. *Organization* 4:355–73.

Curie-Cohen, M., et al. 1983. The effects of dominance on mating behavior and paternity in a captive troop of rhesus monkeys. *American Journal of Primatology* 5:127–38.

Daly, M., and M. Wilson. 1988. *Homicide.* Hawthorne, NY: Aldine de Gruyter.

Damasio, A. R. 1999. *The Feeling of What Happens: Body and Emotion in the Making of Consciousness.* New York: Harcourt.

Davies, N. B. 1992. *Dunnock Behaviour and Social Evolution.* Oxford: Oxford University Press.

Dawkins, R. 1976. *The Selfish Gene.* Oxford: Oxford University Press.

de Beauvoir, S. 1973 (orig. 1949). *The Second Sex.* New York: Vintage Books.

de Waal, F. B. M. 1984. Sex differences in the formation of coalitions among chimpanzees. *Ethology and Sociobiology* 5:239–55.

———. 1986. Integration of dominance and social bonding in primates. *Quarterly Review of Biology* 61:459–79.

———. 1987. Tension regulation and nonreproductive functions of sex in captive bonobos. *National Geographic Research* 3:318–35.

———. 1989. *Peacemaking among Primates,* Cambridge, MA: Harvard University Press.

———. 1993. Sex differences in chimpanzee (and human) behavior: A matter of social values? In *The Origin of Values,* ed. M. Hechter et al., pp. 285–303. New York: Aldine de Gruyter.

———. 1995. Bonobo sex and society. *Scientific American* 272:82–88.

———. 1996a. *Good Natured: The Origins of Right and Wrong in Humans and Other Animals.* Cambridge, MA: Harvard University Press.

———. 1996b. Conflict as negotiation. In *Great Ape Societies,* ed. W. C. McGrew, et al., pp. 159–72. Cambridge, UK: Cambridge University Press.

———. 1997. *Bonobo: The Forgotten Ape.* Berkeley: University of California Press.

———. 1999. The end of nature versus nurture. *Scientific American* 281:94–99.

———. 2000. Primates: A natural heritage of conflict resolution. *Science* 289:586–90.

———. 2000. Survival of the Rapist. *New York Times,* April 2, 2000.

———. 2001. *The Ape and the Sushi Master: Cultural Reflections by a Primatologist.* New York: Basic Books.

———. 2006. *Primates and Philosophers: How Morality Evolved,* ed. S. Macedo and J. Ober. Princeton, NJ: Princeton University Press.

———. 2007 (orig. 1982). *Chimpanzee Politics: Power and Sex among Apes.* Baltimore, MD: Johns Hopkins University Press.

———. 2008. Putting the altruism back into altruism: The evolution of empathy. *Annual Review of Psychology* 59:279–300.

———. 2009. *The Age of Empathy: Nature's Lessons for a Kinder Society.* New York: Harmony.

———. 2013. *The Bonobo and the Atheist: In Search of Humanism among the Primates.* New York: Norton.

———. 2016. *Are We Smart Enough to Know How Smart Animals Are?* New York: Norton.

———. 2019. *Mama's Last Hug: Animal Emotions and What They Tell Us About Ourselves.* New York: Norton.

de Waal, F. B. M., and K. E. Bonnie. 2009. In tune with others: The social side of primate culture. In *The Question of Animal Culture,* ed. K. Laland and G. Galef, pp. 19–39. Cambridge, MA: Harvard University Press.

de Waal, F. B. M., K. Leimgruber, and A. R. Greenberg. 2008. Giving is self-rewarding for monkeys. *Proceedings of the National Academy of Sciences USA* 105:13685–89.

de Waal, F. B. M., and L. M. Luttrell. 1985. The formal hierarchy of rhesus monkeys: An investigation of the bared-teeth display. *American Journal of Primatology* 9:73–85.

de Waal, F. B. M., and J. J. Pokorny. 2008. Faces and behinds: Chimpanzee sex perception. *Advanced Science Letters* 1:99–103.

de Waal, F. B. M., and S. D. Preston. 2017. Mammalian empathy: Behavioral manifestations and neural basis. *Nature Reviews: Neuroscience* 18:498–509.

de Waal, F. B. M., and A. van Roosmalen. 1979. Reconciliation and consolation among chimpanzees. *Behavioral Ecology and Sociobiology* 5:55–66.

Deaner, R. O., S. M. Balish, and M. P. Lombardo. 2015. Sex differences in sports interest and motivation: An evolutionary perspective. *Evolutionary Behavioral Sciences* 10:73–97.

Deardorff, J., et al. 2010. Father absence, body mass index, and pubertal timing in girls: Differential effects by family income and ethnicity. *Journal of Adolescent Health* 48:441–47.

Demuru, E., et al. 2020. Foraging postures are a potential communicative signal in female bonobos. *Scientific Reports* 10:15431.

Demuru, E., P. F. Ferrari, and E. Palagi. 2018. Is birth attendance a uniquely human feature? New evidence suggests that bonobo females protect and support the parturient. *Evolution and Human Behavior* 39:502–10.

Demuth, S., and S. L. Brown. 2004. Family structure, family processes, and adolescent delinquency: The significance of parental absence versus parental gender. *Journal of Research in Crime and Delinquency* 41:58–81.

Denworth, L. 2020. *Friendship: The Evolution, Biology, and Extraordinary Power of Life's Fundamental Bond.* New York: Norton.

Derks, B., et al. 2018. De keuze van vrouwen voor deeltijd is minder vrij dan we denken. *Sociale Vraagstukken,* November 23, 2018 (Dutch).

Derntl, B., et al. 2010. Multidimensional assessment of empathic abilities: Neural correlates and gender differences. *Psychoneuroendocrinology* 35:67–82.

Despret, V. 2009. Culture and gender do not dissolve into how scientists "read" nature: Thelma Rowell's heterodoxy. In *Rebels of Life: Iconoclastic Biologists in the Twentieth Century,* ed. O. Harman and M. Friedrich, pp. 340–55. New Haven, CT: Yale University Press.

Diamond, J. 1992. *The Third Chimpanzee: The Evolution and Future of the Human Animal.* New York: HarperCollins.

Diamond, M., and H. K. Sigmundson. 1997. Sex reassignment at birth: Long-term review and clinical implications. *Archives of Pediatrics and Adolescent Medicine* 151:298–304.

Dienske, H., W. van Vreeswijk, and H. Koning. 1980. Adequate mothering by partially isolated rhesus monkeys after observation of maternal care. *Journal of Abnormal Psychology* 89:489–92.

Diogo, R., J. L. Molnar, and B. Wood. 2017. Bonobo anatomy reveals stasis and mosaicism in chimpanzee evolution, and supports bonobos as the most appropriate extant model for the common ancestor of chimpanzees and humans. *Scientific Reports* 7:608.

DiPietro, J. A. 1981. Rough and tumble play: A function of gender. *Developmental Psychology,* 17:50–58.

Dixon, A. 2010. Homosexual behaviour in primates. In *Animal Homosexuality: A Biosocial Perspective,* ed. A. Poiani, pp. 381–99. Cambridge, UK: Cambridge University Press.

Eckes, T., and H. M. Trautner, eds. 2000. *The Developmental Social Psychology of Gender.* New York: Psychology Press.

Edwards, C. P. 1993. Behavioral sex differences in children of diverse cultures: The case of nurturance to infants. In *Juvenile Primates: Life History, Development, and Behavior,* ed. M. E. Pereira and L. A. Fairbanks, pp. 327–38. New York: Oxford University Press.

Edwards, C. P. 2005. Children's play in cross-cultural perspective: A new look at the six cultures study. *Cross-Cultural Research* 34:318–38.

Ehmann, B., et al. 2021. Sex-specific social learning biases and learning outcomes in wild orangutans. *PLOS* 19: e3001173.

Ellis, B. J., et al. 2003. Does father absence place daughters at special risk for early sexual activity and teenage pregnancy? *Child Development* 74:801–21.

Fagen, R. 1993. Primate juveniles and primate play. In *Primate Juveniles: Life History, Development, and Behavior,* ed. M. E. Pereira and J. A. Fairbanks, pp. 182–96. New York: Oxford University Press.

Fairbanks, L. 2000. Maternal investment throughout the life span in Old World monkeys. In *Old World Monkeys,* ed. P. F. Whitehead and C. J. Jolly, pp. 341–67. Cambridge, UK: Cambridge University Press.

Fairbanks, L. A. 1990. Reciprocal benefits of allomothering for female vervet monkeys. *Animal Behaviour* 40:553–62.

———. 1993. Juvenile vervet monkeys: Establishing relationships and practicing skills for the future. In *Juvenile Primates: Life History, Development, and Behavior,* ed. M. E. Pereira and L. A. Fairbanks, pp. 211–27. New York: Oxford University Press.

Fausto-Sterling, A. 1993. The five sexes: Why male and female are not enough. *The Sciences* 33:20–24.

Fedigan, L. M. 1982. *Primate Paradigms: Sex Roles and Social Bonds.* Montreal: Eden Press.

———. 1994. Science and the successful female: Why there are so many women primatologists. *American Anthropologist* 96:529–40.

Feldblum, J. T., et al. 2014. Sexually coercive male chimpanzees sire more offspring. *Current Biology* 24:2855–60.

Feldman, R., K. Braun, and F. A. Champagne. 2019. The neural mechanisms and consequences of paternal caregiving. *Nature Reviews Neuroscience* 20:205–24.

FeldmanHall, O., et al. 2016. Moral chivalry: Gender and harm sensitivity predict costly altruism. *Social Psychological and Personality Science* 7:542–51.

Finch, C. 2016. Compassionate ostrich offers comfort to baby elephants at orphaned animal sanctuary. *My Modern Met,* October 8, 2016.

Flack, J. C., D. C. Krakauer, and F. B. M. de Waal. 2005. Robustness mechanisms in primate societies: A perturbation study. *Proceedings of the Royal Society London B* 272:1091–99.

Flanagan, J. 1989. Hierarchy in simple "egalitarian" societies. *Annual Review of Anthropology* 18:245–66.

Flemming, A. S., et al. 2002. Mothering begets mothering: The transmission of behavior and its neurobiology across generations. *Pharmacology, Biochemistry and Behavior* 73:61–75.

Flores, A. R., et al. 2016. *How Many Adults Identify as Transgender in the United States?* Los Angeles: UCLA Williams Institute.

Foerster, S., et al. 2016. Chimpanzee females queue but males compete for social status. *Scientific Reports* 6:35404.

Ford, C. S., and F. A. Beach. 1951. *Patterns of Sexual Behavior.* New York: Harper and Brothers.

Forman, J., et al. 2019. Automobile injury trends in the contemporary fleet: Belted occupants in frontal collisions. *Traffic Injury Prevention* 20:607–12.

Forrester, G. S., et al. 2019. The left cradling bias: An evolutionary facilitator of social cognition? *Cortex* 118:116–31.

Foster, M. W., et al. 2009. Alpha male chimpanzee grooming patterns: Implications for dominance "style." *American Journal of Primatology* 71:136–44.

Fraser, O. N., and F. Aureli. 2008. Reconciliation, consolation and postconflict behavioral specificity in chimpanzees. *American Journal of Primatology* 70:1114–23.

French, M. 1985. *Beyond Power: On Women, Men, and Morals.* New York: Ballantine Books.

Frumin, I., et al. 2015. A social chemosignaling function for human handshaking. *eLife* 4:e05154.

Fry, D. P. 2006. *The Human Potential for Peace.* New York: Oxford University Press.

———. 2013. *War, Peace, and Human Nature: The Convergence of Evolutionary and Cultural Views.* Oxford: Oxford University Press.

Fujita, S., and E. Inoue. 2015. Sexual behavior and mating strategies. In *Mahale Chimpanzees: 50 Years of Research,* ed. M. Nakamura et al. Cambridge, UK: Cambridge University Press.

Furuichi, T. 2019. *Bonobo and Chimpanzee: The Lessons of Social Coexistence.* Singapore: Springer Nature.

Furuichi, T., et al. 2014. Why do wild bonobos not use tools like chimpanzees do? *Behaviour* 152:425–60.

Galdikas, B. M. F. 1995. *Reflections of Eden: My Years with the Orangutans of Borneo.* Boston: Little, Brown.

Ganna, A., et al. 2019. Large-scale GWAS reveals insights into the genetic architecture of same-sex sexual behavior. *Science* 365:eaat7693.

Garcia-Falgueras, A., and D. F. Swaab. 2008. A sex difference in the hypothalamic uncinate nucleus: Relationship to gender identity. *Brain* 131:3132–46.

Gavrilets, S., and W. R. Rice. 2006. Genetic models of homosexuality: Generating testable predictions. *Proceedings of the Royal Society B* 273:3031–38.

Ghiselin, M. 1974. *The Economy of Nature and the Evolution of Sex.* Berkeley: University of California Press.

Goldfoot, D. A., et al. 1980. Behavioral and physiological evidence of sexual climax in the female stump-tailed macaque. *Science* 208:1477–79.

Goldhagen, D. J. 1996. *Hitler's Willing Executioners: Ordinary Germans and the Holocaust.* New York: Knopf.

Goldsborough, Z., et al. 2020. Do chimpanzees console a bereaved mother? *Primates* 61:93–102.

Goldstein, J. S. 2001. *War and Gender: How Gender Shapes the War System and Vice Versa.* Cambridge, UK: Cambridge University Press.

Goodall, J. 1979. Life and death at Gombe. *National Geographic* 155:592–621.

———. 1986. *The Chimpanzees of Gombe: Patterns of Behavior.* Cambridge, MA: Belknap Press.

Gould, S. J. 1977. *Ontogeny and Phylogeny.* Cambridge, MA: Belknap Press.

———. 1993. Male nipples and clitoral ripples. *Columbia: Journal of Literature and Art* 20:80–96.

Gowaty, P. A. 1997. Introduction: Darwinian Feminists and Feminist Evolutionists. In *Feminism and Evolutionary Biology,* ed. P. A. Gowaty, pp. 1–17. New York: Chapman and Hall.

Gowaty, P. A., Y.-K. Kim, and W. W. Anderson. 2012. No evidence of sexual selection in a repetition of Bateman's classic study of *Drosophila melanogaster. Proceedings of the National Aacademy of Sciences USA* 109:11740–45.

Grammer, K., L. Renninger, and B. Fischer. 2005. Disco clothing, female sexual motivation, and relationship status: Is she dressed to impress? *Journal of Sex Research* 41:66–74.

Grawunder, S., et al. 2018. Higher fundamental frequency in bonobos is explained by larynx morphology. *Current Biology* 28:R1188–89.

Gray, J. 1992. *Men Are from Mars, Women Are from Venus: A Practical Guide for Improving Communication and Getting What You Want in Your Relationships.* New York: HarperCollins.

Greenberg, D. 1988. *The Construction of Homosexuality.* Chicago: University of Chicago Press.

Gülgöz, S., et al. 2019. Similarity in transgender and cisgender children's gender development. *Proceedings of the National Academy of Sciences USA* 116:24480–85.

Gutmann, M. C. 1997. Trafficking in men: The anthropology of masculinity. *Annual Review of Anthropology* 26:385–409.

Haig, D. 2004. The inexorable rise of gender and the decline of sex: Social change in academic titles, 1945–2001. *Archives of Sexual Behavior* 33:87–96.

Hall, J. A. 2011. Sex differences in friendship expectations: A meta-analysis. *Journal of Social and Personal Relationships* 28:723–47.

Hall, K. R. L., and I. DeVore. 1965. Baboon social behavior. In *Primate Behavior: Field Studies of Monkeys and Apes,* ed. I. DeVore, pp. 53–110. New York: Holt, Rinehart and Winston.

Hallal, P. C., et al. 2012. Global physical activity levels: Surveillance progress, pitfalls, and prospects. *Lancet* 380:247–57.

Halley, J. E. 1994. Sexual orientation and the politics of biology: A critique of the argument from immutability. *Stanford Law Review* 46:503–68.

Haraway, D. 1989. *Primate Visions: Gender, Race, and Nature in the World of Modern Science.* New York: Routledge.

Harlan, R. 1827. Description of a hermaphrodite orang outang. *Proceedings of the Academy of Natural Sciences Philadelphia* 5:229–36.

Harris, J. R. 1998. *The Nurture Assumption: Why Children Turn Out the Way They Do.* London: Bloomsbury.

Harrison, J., et al. 2011. Belzec, Sobibor, Treblinka: Holocaust Denial and Operation Reinhard. *Holocaust Controversies,* http://holocaustcontroversies .blogspot.com/2011/12/belzec-sobibor-treblinka-holocaust.html.

Haselton, M. G., et al. 2007. Ovulatory shifts in human female ornamentation: Near ovulation, women dress to impress. *Hormones and Behavior* 51:40–45.

Hassett, J. M., E. R. Siebert, and K. Wallen. 2008. Sex differences in rhesus monkey toy preferences parallel those of children. *Hormones and Behavior* 54:359–64.

Hawkes, K., and J. E. Coxworth. 2013. Grandmothers and the evolution of human longevity: A review of findings and future directions. *Evolutionary Anthropology* 22:294–302.

Hayes, C. 1951. *The Ape in Our House.* New York: Harper.

Hecht, E. E., et al. 2021. Sex differences in the brains of capuchin monkeys. *Journal of Comparative Neurology* 2:327–39.

Henrich, J., and F. J. Gil-White. 2001. The evolution of prestige: Freely conferred deference as a mechanism for enhancing the benefits of cultural transmission. *Evolution and Human Behavior* 22:165–96.

Herman, R. A., M. A. Measday, and K. Wallen. 2003. Sex differences in interest in infants in juvenile rhesus monkeys: Relationship to prenatal androgen. *Hormones and Behavior* 43:573–83.

Herschberger, R. 1948. *Adam's Rib.* New York: Harper and Row.

Hesse, M. 2019. Elizabeth Holmes's weird, possibly fake baritone is actually her least baffling quality. *Washington Post,* March 21, 2019.

Hill, S. E., R. P. Proffitt Levya, and D. J. DelPriore. 2016. Absent fathers and sexual strategies. *Psychologist* 29:436–39.

Hines, M. 2011. Gender development and the human brain. *Annual Review of Neuroscience* 34:69–88.

Hockings, K. J., J. R. Anderson, and T. Matsuzawa. 2006. Road crossing in chimpanzees: A risky business. *Current Biology* 16:668–70.

Hockings, K. J., et al. 2007. Chimpanzees share forbidden fruit. *PLoS ONE* 9:e886.

Hohmann, G., and B. Fruth. 2011. Is blood thicker than water? In *Among African Apes,* ed. M. M. Robbins and C. Boesch, pp. 61–76. Berkeley: University of California Press.

Hopkins, W. D. 2004. Laterality in maternal cradling and infant positional biases: Implications for the development and evolution of hand preferences in nonhuman primates. *International Journal of Primatology* 25:1243–65.

Hopkins, W. D., and M. de Lathouwers. 2006. Left nipple preferences in infant *Pan paniscus* and *P. troglodytes. International Journal of Primatology* 27:1653–62.

Hoquet, T., et al. 2020. Bateman's data: Inconsistent with "Bateman's Principles." *Ecology and Evolution* 10:10325–42.

Horner, V., and F. B. M. de Waal. 2009. Controlled studies of chimpanzee cultural transmission. *Progress in Brain Research* 178:3–15.

Horner, V., D. J. Carter, M. Suchak, and F. B. M. de Waal. 2011. Spontaneous prosocial choice by chimpanzees. *Proceedings of the Academy of Sciences USA* 108:13847–51.

Horner, V., et al. 2010. Prestige affects cultural learning in chimpanzees. *PLoS ONE* 5:e10625.

Hrdy, S. B. 1977. *The Langurs of Abu: Female and Male Strategies of Reproduction.* Cambridge, MA: Harvard University Press.

———. 1981. *The Woman That Never Evolved.* Cambridge, MA: Harvard University Press.

———. 1999. *Mother Nature: A History of Mothers, Infants, and Natural Selection.* New York: Pantheon.

———. 2000. The optimal number of fathers: Evolution, demography, and history in the shaping of female mate preferences. *Annals of the New York Academy of Sciences* 907:75–96.

———. 2009. *Mothers and Others: The Evolutionary Origins of Mutual Understanding.* Cambridge, MA: Belknap Press.

Hyde, J. S., and J. DeLamater. 1997. *Understanding Human Sexuality.* New York: McGraw-Hill.

Hyde, J. S., et al. 2008. Gender similarities characterize math performance. *Science* 321:494–95.

Idani, G. 1990. Relations between unit-groups of bonobos at Wamba, Zaire: Encounters and temporary fusions. *African Study Monographs* 11:153–86.

———. 1993. A bonobo orphan who became a member of the wild group. *Primate Research* 9:97–105.

Jabbour, J., et al. 2020. Robust evidence for bisexual orientation among men. *Proceedings of the National Academy of Sciences USA* 117:18369–77.

Jadva, V., M. Hines, and S. Golombok. 2010. Infants' preferences for toys, colors, and shapes: Sex differences and similarities. *Archives of Sexual Behavior* 39:1261–73.

Jannini, E. A., O. Buisson, and A. Rubio-Casillas. 2014. Beyond the G-spot: Clitourethrovaginal complex anatomy in female orgasm. *Nature Reviews Urology* 11:531–38.

Jolly, A. 1999. *Lucy's Legacy: Sex and Intelligence in Human Evolution.* Cambridge, MA: Harvard University Press.

Jones, A. 2002. Gender and genocide in Rwanda. *Journal of Genocide Research* 4:65–94.

Jones, L. K., B. M. Jennings, M. Higgins, and F. B. M. de Waal. 2018. Ethological observations of social behavior in the operating room. *Proceedings of the National Academy of Sciences USA* 115:7575–80.

Jordan-Young, R. M., and K. Karkazis. 2019. *Testosterone: An Unauthorized Biography.* Cambridge, MA: Harvard University Press.

Kahlenberg, S. M., and R. W. Wrangham. 2010. Sex differences in chimpanzees' use of sticks as play objects resemble those of children. *Current Biology* 20:R1067–68.

Kahneman, D. 2013. *Thinking, Fast and Slow.* New York: Farrar, Straus and Giroux.

Kano, T. 1992. *The Last Ape: Pygmy Chimpanzee Behavior and Ecology.* Stanford, CA: Stanford University Press.

———. 1998. Comments on C. B. Stanford. *Current Anthropology* 39:410–11.

Killen, M., and E. Turiel. 1991. Conflict resolution in preschool social interactions. *Early Education and Development* 2:240–55.

Kinsey, A. C., W. R. Pomeroy, and C. E. Martin. 1948. *Sexual Behavior in the Human Male.* Philadelphia: Saunders.

Kirkpatrick, M. 1987. Clinical implications of lesbian mother studies. *Journal of Homosexuality* 14:201–11.

Klofstad, C. A., R. C. Anderson, and S. Peters. 2012. Sounds like a winner: Voice pitch influences perception of leadership capacity in both men and women. *Proceedings of the Royal Society B* 279:2698–704.

Knott, C. D., and S. Kahlenberg. 2007. Orangutans in perspective: Forced copulations and female mating resistance. In *Primates in Perspective,* ed. S. Bearder et al., pp. 290–305. New York: Oxford University Press.

Köhler, W. 1925. *The Mentality of Apes.* New York: Vintage.

Konner, M. J. 1976. Maternal care, infant behavior, and development among the !Kung. In *Kalahari Hunter Gatherers,* ed. R. B. Lee and I. DeVore, pp. 218–45. Cambridge, MA: Harvard University Press.

Konner, M. J. 2015. *Women After All: Sex, Evolution, and the End of Male Supremacy.* New York: Norton.

Koski, S. E., K. Koops, and E. H. M. Sterck. 2007. Reconciliation, relationship quality, and postconflict anxiety: Testing the integrated hypothesis in captive chimpanzees. *American Journal of Primatology* 69:158–72.

Kret, M. E., and M. Tomonaga. 2016. Getting to the bottom of face processing: Species-specific inversion effects for faces and behinds in humans and chimpanzees (*Pan troglodytes*). *PLoS ONE* 11:e0165357.

Krupenye, C., et al. 2016. Great apes anticipate that other individuals will act according to false beliefs. *Science* 354:110–14.

Kummer, H. 1971. *Primate Societies: Group Techniques of Ecological Adaptation.* Chicago: Aldine.

———. 1995. *In Quest of the Sacred Baboon: A Scientist's Journey.* Princeton, NJ: Princeton University Press.

Lafreniere, P. 2011. Evolutionary functions of social play: Life histories, sex differences, and emotion regulation. *American Journal of Play* 3:464–88.

Lagerspetz, K. M., et al. 1988. Is indirect aggression typical of females? *Aggressive Behavior* 14:403–14.

Lamb, M. E., and D. Oppenheim. 1989. Fatherhood and father-child relationships: Five years of research. In *Fathers and Their Families,* ed. S. H. Cath et al., pp. 11–26. Hillsdale, NJ: Analytic Press.

Lancaster, J. B. 1971. Play-mothering: The relations between juvenile females and young infants among free-ranging vervet monkeys (*Cercopithecus aethiops*). *Folia primatologica* 15:161–82.

Långström, N., et al. 2010. Genetic and environmental effects on same-sex sexual behavior: A population study of twins in Sweden. *Archives of Sexual Behavior* 39:75–80.

Lappan, S. 2008. Male care of infants in a siamang population including socially monogamous and polyandrous groups. *Behavioral Ecology and Sociobiology* 62:1307–17.

Laqueur, T. W. 1990. *Making Sex: Body and Gender from the Greeks to Freud.* Cambridge, MA: Harvard University Press.

Leavitt, H. J. 2003. Why hierarchies thrive. *Harvard Business Review,* March 2003.

Leca, J.-B., N. Gunst, and P. L. Vasey. 2014. Male homosexual behavior in a free-ranging all-male group of Japanese macaques at Minoo, Japan. *Archives of Sexual Behavior* 43:853–61.

Lemaître, J.-F., et al. 2020. Sex differences in adult lifespan and aging rates of mortality across wild mammals. *Proceedings of the National Academy of Sciences USA* 117:8546–53.

Lerner, R. M. 1978. Nature, nurture, and dynamic interactionism. *Human Development* 21:1–20.

Lethmate, J., and G. Dücker. 1973. Untersuchungen zum Selbsterkennen im Spiegel bei Orang-Utans und einigen anderen Affenarten. *Zeitschrift für Tierpsychologie* 33:248–69 (German).

LeVay, S. 1991. A difference in hypothalamic structure between homosexual and heterosexual men. *Science* 253:1034–37.

———. 1996. *Queer Science: The Use and Abuse of Research into Homosexuality.* Cambridge, MA: MIT Press.

Lever, J. 1976. Sex differences in the games children play. *Social Problems* 23:478–87.

Lévi-Strauss, C. 1969 (orig. 1949). *The Elementary Structures of Kinship.* Boston: Beacon Press.

Lewis, R. J. 2018. Female power in primates and the phenomenon of female dominance. *Annual Review of Anthropology* 47:533–51.

Leyk, D., et al. 2007. Hand-grip strength of young men, women and highly trained female athletes. *European Journal of Applied Physiology* 99:415–21.

Lindegaard, M. R., et al. 2017. Consolation in the aftermath of robberies resembles post-aggression consolation in chimpanzees. *PLoS ONE* 12:e0177725.

Linden, E. 2002. The wife beaters of Kibale. *Time* 160:56–57.

Lindenfors, P., J. L. Gittleman, and K. E. Jones. 2007. Sexual size dimorphism in mammals. In *Evolutionary Studies of Sexual Size Dimorphism,* ed. D. J. Fairbairn, W. U. Blanckenhorn, and T. Szekely, pp. 16–26. Oxford: Oxford University Press.

Lloyd, E. A. 2005. *The Case of the Female Orgasm: Bias in the Science of Evolution.* Cambridge, MA: Harvard University Press.

Lonsdorf, E. V., L. E. Eberly, and A. E. Pusey. 2004. Sex differences in learning in chimpanzees. *Nature* 428:715–16.

Lonstein, J. S., and G. J. de Vries. 2000. Sex differences in the parental behavior of rodents. *Neuroscience and Biobehavioral Reviews* 24:669–86.

Losin, E. A., et al. 2012. Own-gender imitation activates the brain's reward circuitry. *Social Cognitive and Affective Neuroscience* 7:804–10.

Ludwig, A. M. 2002. *King of the Mountain: The Nature of Political Leadership.* Lexington: University Press of Kentucky.

Luef, E. M., T. Breuer, and S. Pika. 2016. Food-associated calling in gorillas (*Gorilla g. gorilla*) in the wild. *PLoS ONE* 11:e0144197.

Lundström, J. N., et al. 2013. Maternal status regulates cortical responses to the body odor of newborns. *Frontiers in Psychology* 4:597.

Lutchmaya, S., and S. Baron-Cohen. 2002. Human sex differences in social and non-social looking preferences, at 12 months of age. *Infant Behavior and Development* 25:319–25.

Maccoby, E. E. 1998. *The Two Sexes: Growing up Apart, Coming Together.* Cambridge, MA: Belknap Press.

MacDonald, K., and R. D. Parke. 1986. Parent-child physical play: The effects of sex and age of children and parents. *Sex Roles* 15:367–78.

Maerker, A. 2005. Scenes from the museum: The hermaphrodite monkey and stage management at La Specola. *Endeavour* 29:104–8.

Maestripieri, D., and S. Pelka. 2002. Sex differences in interest in infants across the lifespan: A biological adaptation for parenting? *Human Nature* 13:327–44.

Maggioncalda, A. N., N. M. Czekala, and R. M. Sapolsky. 2002. Male orang-utan subadulthood: A new twist on the relationship between chronic stress and developmental arrest. *American Journal of Physical Anthropology* 118:25–32.

Maglaty, J. 2011. When did girls start wearing pink? Smithsonian, April 7, 2011.

Mann, D. 2017. *Become the Alpha Male: How to Be an Alpha Male, Dominate in Both the Boardroom and Bedroom, and Live the Life of a Complete Badass.* Independently published.

Maple, T. 1980. *Orangutan Behavior.* New York: Van Nostrand Reinhold.

Marshall, P., A. Bartolacci, and D. Burke. 2020. Human face tilt is a dynamic social signal that affects perceptions of dimorphism, attractiveness, and dominance. *Evolutionary Psychology* 18:1–15.

Martin, C. L., and R. A. Fabes. 2001. The stability and consequences of young children's same-sex peer interactions. *Developmental Psychology* 37:431–46.

Martin, R. D. 2019. No substitute for sex: "Gender" and "sex" have very different meanings. *Psychology Today,* August 20, 2019.

Maslow, A. 1936. The role of dominance in the social and sexual behavior of infra-human primates. *Journal of Genetic Psychology* 48:261–338 and 49:161–98.

Massen, J. J. M., et al. 2010. Generous leaders and selfish underdogs: Prosociality in despotic macaques. *PLoS ONE* 5:e9734.

Mast, M. S. 2002. Female dominance hierarchies: Are they any different from males'? *Personality and Social Psychology Bulletin* 28:29–39.

———. 2004. Men are hierarchical, women are egalitarian: An implicit gender stereotype. *Swiss Journal of Psychology* 62:107–11.

Matevia, M. L., F. G. P. Patterson, and W. A. Hillix. 2002. Pretend play in a signing gorilla. In *Pretending and Imagination in Animals and Children,* ed. R. W. Mitchell, pp. 285–306. Cambridge, UK: Cambridge University Press.

Matsuzawa, T. 1997. The death of an infant chimpanzee at Bossou, Guinea. *Pan Africa News* 4:4–6.

Mayhew, R. 2004. *The Female in Aristotle's Biology: Reason or Rationalization.* Chicago: University of Chicago Press.

Mayr, E. 1982. *The Growth of Biological Thought.* Cambridge, MA: Harvard University Press.

McAlone, N. 2015. Here's how Janet Jackson's infamous "nipplegate" inspired the creation of YouTube. *Business Insider,* October 3, 2015.

McBee, T. P. 2016. Until I was a man, I had no idea how good men had it at work. *Quartz,* May 13, 2016.

McCann, S. J. H. 2001. Height, social threat, and victory margin in presidential elections (1894–1992). *Psychological Reports* 88:741–42.

McCarthy, M. M. 2016. Multifaceted origins of sex differences in the brain. *Philosophical Transactions of the Royal Society B* 371:20150106.

McCloskey, D. N. 1999. *Crossing: A Memoir.* Chicago: University of Chicago Press.

McElwain, G. S. 2020. *Mary Midgley: An Introduction.* London: Bloomsbury.

McGrew, W. C. 1992. *Chimpanzee Material Culture.* Cambridge, UK: Cambridge University Press.

McGrew, W. C., and L. F. Marchant. 1998. Chimpanzee wears a knotted skin "necklace." *Pan African News* 5:8–9.

Mead, M. 2001 (orig. 1949). *Male and Female.* New York: Perennial.

Merkle, S. 1989. Sexual differences as adaptation to the different gender roles in the frog *Xenopus laevis* Daudin. *Journal of Comparative Physiology B* 159:473–80.

Meston, C. M., and D. M. Buss. 2007. Why humans have sex. *Archives of Sexual Behavior* 36:477–507.

Meyer-Bahlburg, H. F. L. 2005. Gender identity outcome in female-raised 46,XY persons with penile agenesis, cloacal exstrophy of the bladder, or penile ablation. *Archives of Sexual Behavior* 34:423–38.

Michele, A., and T. Fisher. 2003. Truth and consequences: Using the bogus pipeline to examine sex differences in self-reported sexuality. *Journal of Sex Research* 40:27–35.

Midgley, M. 1995. *Beast and Man: The Roots of Human Nature.* London: Routledge.

———. 2010. *The Solitary Self: Darwin and the Selfish Gene.* Durham, UK: Acumen.

Mitani, J. C., and T. Nishida. 1993. Contexts and social correlates of long-distance calling by male chimpanzees. *Animal Behaviour* 45:735–46.

Mitchell, R. W., ed. 2002. *Pretending and Imagination in Animals and Children.* Cambridge, UK: Cambridge University Press.

Money, J., J. G. Hampson, and J. Hampson. 1955. An examination of some basic sexual concepts: The evidence of human hermaphroditism. *Bulletin of Johns Hopkins Hospital* 97:301–19.

Montagu, M. F. A. 1962. *The Natural Superiority of Women.* New York: Macmillan.

———, ed. 1973. *Man and Aggression.* New York: Oxford University Press.

Morgan, M. H., and D. R. Carrier. 2013. Protective buttressing of the human fist and the evolution of hominin hands. *Journal of Experimental Biology* 216:236–44.

Morris, D. 1977. *Manwatching: A Field Guide to Human Behavior.* London: Jonathan Cape.

———. 2017 (orig. 1967). *The Naked Ape: A Zoologist's Study of the Human Animal.* London: Penguin.

Morris, J. 1974. *Conundrum.* New York: New York Review of Books,

Morrison, T. 2019. Goodness. *New York Times Book Review,* September 8, 2019, pp. 16–17.

Moscovice, L. R., et al. 2019. The cooperative sex: Sexual interactions among female bonobos are linked to increases in oxytocin, proximity, and coalitions. *Hormones and Behavior* 116:104581.

Moye, D. 2019. Speech coach has a theory on Theranos CEO Elizabeth Holmes and her deep voice. *Huffington Post,* April 11, 2019.

Muller, M. N., et al. 2011. Sexual coercion by male chimpanzees shows that female choice may be more apparent than real. *Behavioral Ecology and Sociobiology* 65:921–33.

Muller, M. N., S. M. Kahlenberg, and R. W. Wrangham. 2009. Male Aggression against females and sexual coercion in chimpanzees. In *Sexual Coercion in Primates and Humans: An Evolutionary Perspective on Male Aggression Against Females,* ed. M. N. Muller and R. W. Wrangham, pp. 184–217. Cambridge, MA: Harvard University Press.

Murphy, S. M. 1992. *A Delicate Dance: Sexuality, Celibacy, and Relationships Among Catholic Clergy and Religious.* New York: Crossroad.

Murray, C. M., E. Wroblewski, and A. E. Pusey. 2007. New case of intragroup infanticide in the chimpanzees of Gombe National Park. *International Journal of Primatology* 28:23–37.

Musgrave, S., et al. 2016. Tool transfers are a form of teaching among chimpanzees. *Scientific Reports* 6:34783.

Musgrave, S., et al. 2020. Teaching varies with task complexity in wild chimpanzees. *Proceedings of the National Academy of Sciences USA* 117:969–76.

Nadler, R. D., et al. 1985. Serum levels of gonadotropins and gonadal steroids, including testosterone, during the menstrual cycle of the chimpanzee. *American Journal of Primatology* 9:273–84.

Nash, R., et al. 2006. Cosmetics: They influence more than Caucasian female facial attractiveness. *Journal of Applied Social Psychology* 36:493–504.

Nelson, A. 2005. Children's toy collections in Sweden: A less gender-typed country? *Sex Roles* 52:93–102.

Nguyen, N., R. C. van Horn, S. C. Alberts, and J. Altmann. 2009. "Friendships" between new mothers and adult males: Adaptive benefits and determinants in wild baboons (*Papio cynocephalus*). *Behavioral Ecology and Sociobiology* 63:1331–44.

Nicholls, H. 2014. In conversation with Jane Goodall. *Mosaic Science,* March 31, 2014, mosaicscience.com/story/conversation-with-jane-goodall.

Nieuwenhuijsen, K. 1985. *Geslachtshormonen en Gedrag bij de Beermakaak.* Ph.D. thesis, Erasmus University, Rotterdam (Dutch).

Nishida, T. 1996. The death of Ntologi: The unparalleled leader of M Group. *Pan Africa News* 3:4.

Nishida, T. 2012. *Chimpanzees of the Lakeshore.* Cambridge, UK: Cambridge University Press.

Nishida, T., and K. Hosaka. 1996. Coalition strategies among adult male chimpanzees of the Mahale Mountains, Tanzania. In *Great Ape Societies,* ed. W. C. McGrew et al., pp. 114–34. Cambridge, UK: Cambridge University Press.

Nolen-Hoeksema, S., B. E. Wisco, and S. Lyubomirsky. 2008. Rethinking rumination. *Perspectives on Psychological Science* 3:400–24.

Nussbaum, M. 2001. *Upheavals of Thought: The Intelligence of Emotions.* Cambridge, UK: Cambridge University Press.

O'Connell, C. 2015. *Elephant Don: The Politics of a Pachyderm Posse.* Chicago: University of Chicago Press.

O'Connell, H. E., K. V. Sanjeevan, and J. M. Hutson. 2005. Anatomy of the clitoris. *Journal of Urology* 174:1189–95.

O'Connell, M. 2017. *To Be a Machine.* London: Granta.

O'Toole, A. J., et al. 1998. The perception of face gender: The role of stimulus structure in recognition and classification. *Memory and Cognition* 26:146–60.

O'Toole, A. J., J. Peterson, and K. A. Deffenbacher. 1996. An "other-race effect" for classifying faces by gender. *Perception* 25:669–76.

Oakley, K. 1950. *Man the Tool Maker.* London: Trustees of the British Museum.

Orbach, D., and P. Brennan. 2019. Functional morphology of the dolphin clitoris. Presented at Experimental Biology Conference, Orlando, FL.

Ortiz, A. 2020. Diego, the tortoise whose high sex drive helped save his species, retires. *New York Times,* January 12, 2020.

Palagi, E., and E. Demuru. 2017. *Pan paniscus* or *Pan ludens*? Bonobos, playful attitude and social tolerance. In *Bonobos: Unique in Mind and Behavior,* ed. B. Hare and S. Yamamoto, pp. 65–77. Oxford: Oxford University Press.

Palagi, E., et al. 2020. Mirror replication of sexual facial expressions increases the success of sexual contacts in bonobos. *Scientific Reports* 10:18979.

Palagi, E., T. Paoli, and S. Borgognini. 2004. Reconciliation and consolation in captive bonobos (*Pan paniscus*). *American Journal of Primatology* 62:15–30.

Paresky, P. B. 2019. What's the problem with "traditional masculinity"? The frenzy about the APA guidelines has died down. What have we learned? *Psychology Today,* March 10, 2019.

Parish, A. R. 1993. Sex and food control in the "uncommon chimpanzee": How bonobo females overcome a phylogenetic legacy of male dominance. *Ethology and Sociobiology* 15:157–79.

Parish, A. R., and F. B. M. de Waal. 2000. The other "closest living relative": How bonobos (*Pan paniscus*) challenge traditional assumptions about females, dominance, intra- and inter-sexual interactions, and hominid evolution. *Annals of the New York Academy of Sciences* 907:97–113.

Patterson, N., et al. 2006. Genetic evidence for complex speciation of humans and chimpanzees. *Nature* 441:1103–8.

Pauls, R. N. 2015. Anatomy of the clitoris and the female sexual response. *Clinical Anatomy* 28:376–84.

Peirce, L. P. 1993. *The Imperial Harem: Women and Sovereignty in the Ottoman Empire.* Oxford: Oxford University Press.

Pellegrini, A. D. 1989. Elementary school children's rough-and-tumble play. *Early Childhood Research Quarterly* 4:245–60.

Pellegrini, A. D. 2010. The role of physical activity in the development and function of human juveniles' sex segregation. *Behaviour* 147:1633–56.

Pellegrini, A. D., and P. K. Smith. 1998. Physical activity play: The nature and function of a neglected aspect of play. *Child Development* 69:577–98.

Perry, S. 2008. *Manipulative Monkeys: The Capuchins of Lomas Barbudal.* Cambridge, MA: Harvard University Press.

———. 2009. Conformism in the food processing techniques of white-faced capuchin monkeys (*Cebus capucinus*). *Animal Cognition* 12:705–16.

Petr, M., S. Pääbo, J. Kelso, and B. Vernot. 2019. Limits of long-term selection against Neanderthal introgression. *Proceedings of the National Academy of Sciences USA* 116:1639–44.

Pincemy, G., F. S. Dobson, and P. Jouventin. 2010. Homosexual mating displays in penguins. *Ethology* 116:1210–16.

Pinker, S. 2011. *The Better Angels of Our Nature: Why Violence Has Declined.* New York: Viking.

Ploog, D. W., and P. D. MacLean. 1963. Display of penile erection in squirrel monkey (*Saimiri sciureus*). *Animal Behaviour* 32:33–39.

Potts, M., and R. Short. 1999. *Ever Since Adam and Eve: The Evolution of Human Sexuality.* Cambridge, UK: Cambridge University Press.

Prause, N., et al. 2016. Clitorally stimulated orgasms are associated with better control of sexual desire, and not associated with depression or anxiety, compared with vaginally stimulated orgasms. *Journal of Sexual Medicine* 13:1676–85.

Price, D. 2018. Gender socialization is real (complex). *Devon Price,* November 5, 2018 medium.com/@devonprice/gender-socialization-is-real-complex -348f56146925.

Pruetz, J. D. 2011. Targeted helping by a wild adolescent male chimpanzee (*Pan troglodytes verus*): Evidence for empathy? *Journal of Ethology* 29:365–68.

Prüfer, K., et al. 2012. The bonobo genome compared with the chimpanzee and human genomes. *Nature* 486:527–31.

Prum, R. O. 2015. The role of sexual autonomy in evolution by mate choice. In *Current Perspectives on Sexual Selection: What's Left after Darwin?* ed. T. Hoquet, pp. 237–62. Dordrecht: Springer.

Prum, R. O. 2017. *The Evolution of Beauty: How Darwin's Forgotten Theory of Mate Choice Shapes the Animal World.* New York: Doubleday.

Puppo, V. 2013. Anatomy and physiology of the clitoris, vestibular bulbs, and labia minora with a review of the female orgasm and the prevention of female sexual dysfunction. *Clinical Anatomy* 26:134–52.

Pusey, A. E. 1980. Inbreeding avoidance in chimpanzees. *Animal Behaviour* 28:543–52.

Puts, D. A., C. R. Hodges, R. A. Cárdenas, and S. J. C. Gaulin. 2007. Men's voices as dominance signals: Vocal fundamental and formant frequencies influence dominance attributions among men. *Evolution and Human Behavior* 28:340–44.

Reddy, R. B., and J. C. Mitani. 2019. Social relationships and caregiving behavior between recently orphaned chimpanzee siblings. *Primates* 60:389–400.

Reeve, E. 2013. Male pundits fear the natural selection of Fox's female breadwinners. *Atlantic,* May 30, 2013.

Regan, B. C., et al. 2001. Fruits, foliage and the evolution of primate colour vision. *Philosophical Transactions of the Royal Society B* 356:229–83.

Regitz-Zagrosek, V. 2012. Sex and gender differences in health. *EMBO Reports* 13:596–603.

Reinhardt, V., et al. 1986. Altruistic interference shown by the alpha-female of a captive troop of rhesus monkeys. *Folia primatologica* 46:44–50.

Reynolds, V. 1967. *The Apes.* New York: Dutton.

Riley, C. 2019. How to play Patriarchy Chicken: Why I refuse to move out of the way for men. *New Statesman,* February 22, 2019.

Rilling, J. K., and J. S. Mascaro. 2017. The neurobiology of fatherhood. *Current Opinion in Psychology* 15:26–32.

Rippon, G. 2019. *The Gendered Brain: The New Neuroscience that Shatters the Myth of the Female Brain.* New York: Random House.

Robarchek, C. A. 1997. A community of interests: Semai conflict resolution. In *Cultural Variation in Conflict Resolution: Alternatives to Violence,* ed. D. P. Fry and K. Björkqvist, pp. 51–58. Mahwah, NJ: Erlbaum.

Roberts, W. P., and M. Krause. 2002. Pretending culture: Social and cognitive features of pretense in apes and humans. In *Pretending and Imagination in Animals and Children,* ed. R. W. Mitchell, pp. 269–79. Cambridge, UK: Cambridge University Press.

Romans, S., et al. 2003. Age of menarche: The role of some psychosocial factors. *Psychological Medicine* 33:933–39.

Romero, M. T., M. A. Castellanos, and F. B. M. de Waal. 2010. Consolation as possible expression of sympathetic concern among chimpanzees. *Proceedings of the National Academy of Sciences USA* 107:12110–15.

Rosati, A. G., et al. 2020. Social selectivity in aging wild chimpanzees. *Science* 370:473–76.

Rose, A. J., and K. D. Rudolph. 2006. A review of sex differences in peer relationship processes: Potential trade-offs for the emotional and behavioral development of girls and boys. *Psychological Bulletin* 132:98–131.

Roselli, C. E., et al. 2004. The volume of a sexually dimorphic nucleus in the ovine medial preoptic area/anterior hypothalamus varies with sexual partner preference. *Endocrinology* 145:478–83.

Roseth, C. 2018. Children's peacekeeping and peacemaking. In *Peace Ethology: Behavioral Processes and Systems of Peace,* ed. P. Verbeek and B. A. Peters, pp.113–32. Hoboken, NJ: Wiley.

Roughgarden, J. 2004. Review of "Evolution, Gender, and Rape." *Ethology* 110:76.

———. 2017. Homosexuality and evolution: A critical appraisal. In *On Human*

Nature: Biology, Psychology, Ethics, Politics, and Religion, ed. M. Tibayrenc and F. J. Ayala, pp. 495–516. New York: Academic Press.

Rowell, T. E. 1974. *The Social Behavior of Monkeys.* New York: Penguin.

Rubin, Z. 1980. *Children's Friendships.* Cambridge, MA: Harvard University Press.

Rueckert, L., et al. 2011. Are gender differences in empathy due to differences in emotional reactivity? *Psychology* 2:574–78.

Rupp, H. A., and K. Wallen. 2008. Sex differences in response to visual sexual stimuli: A review. *Archives of Sexual Behavior* 37:206–18.

Russell, D. G. D., et al. 2012. Dr. George Murray Levick (1876–1956): Unpublished notes on the sexual habits of the Adélie penguin. *Polar Record* 48:387–93.

Russell, N. 2019. The Nazi's pursuit for a "humane" method of killing. In *Understanding Willing Participants,* vol. 2. Cham, Switzerland: Palgrave Macmillan.

Russell, R. 2009. A sex difference in facial contrast and its exaggeration by cosmetics. *Perception* 38:1211–19.

Rutherford, A. 2018. *Humanimal: How* Homo sapiens *Became Nature's Most Paradoxical Creature.* New York: Experiment.

Rutherford, A. 2020. *How to Argue with a Racist: What Our Genes Do (and Don't) Say About Human Difference.* New York: Experiment.

Safdar, S., et al. 2009. Variations of emotional display rules within and across cultures: A comparison between Canada, USA, and Japan. *Canadian Journal of Behavioural Science* 41:1–10.

Salerno, J., and L. C. Peter-Hagene. 2015. One angry woman: Anger expression increases influence for men, but decreases influence for women, during group deliberation. *Law and Human Behavior* 39:581–92.

Sandel, A. A., K. E. Langergraber, and J. C. Mitani. 2020. Adolescent male chimpanzees (*Pan troglodytes*) form social bonds with their brothers and others during the transition to adulthood. *American Journal of Primatology* 82:e23091.

Sapolsky, R. 1994. *Why Zebras Don't Get Ulcers: A Guide to Stress, Stress-Related Diseases and Coping.* New York: W. H. Freeman.

Sapolsky, R. M. 1997. *The Trouble with Testosterone.* New York: Scribner.

Sauver, J. L. S., et al. 2004. Early life risk factors for Attention-Deficit/Hyperactivity Disorder: A population-based cohort study. *Mayo Clinic Proceedings* 79:1124–31.

Savage-Rumbaugh, S., and B. Wilkerson. 1978. Socio-sexual behavior in *Pan paniscus* and *Pan troglodytes:* A comparative study. *Journal of Human Evolution* 7:327–44.

Savic, I., and P. Lindström. 2008. PET and MRI show differences in cerebral asymmetry and functional connectivity between homo- and heterosexual subjects. *Proceedings of the National Academy of Sciences USA* 105:9403–8.

Savic, I., H. Berglund, and P. Lindström. 2005. Brain response to putative pheromones in homosexual men. *Proceedings of the National Academy of Sciences USA* 102:7356–61.

Savin-Williams, R. C., and Z. Vrangalova. 2013. Mostly heterosexual as a distinct sexual orientation group: A systematic review of the empirical evidence. *Developmental Review* 33:58–88.

Schenkel, R. 1947. Ausdrucks-Studien an Wölfen: Gefangenschafts-Beobachtungen. *Behaviour* 1:81–129 (German).

Schmitt, D. P. 2015. Are women more emotional than men? *Psychology Today,* April 10, 2015.

Schneiderman, I., et al. 2012. Oxytocin during the initial stages of romantic attachment: Relations to couples' interactive reciprocity. *Psychoneuroendocrinology* 37:1277–85.

Schoppe-Sullivan, S. J., et al. 2021. Fathers' parenting and coparenting behavior in dual-earner families: Contributions of traditional masculinity, father nurturing role beliefs, and maternal gate closing. *Psychology of Men and Masculinities,* advance online at doi.org/10.1037/men0000336.

Schulte-Rüther, M., et al. 2008. Gender differences in brain networks supporting empathy. *NeuroImage* 42:393–403.

Schwartz, S. H., and T. Rubel. 2005. Sex differences in value priorities: Cross-cultural and multimethod studies. *Journal of Personality and Social Psychology* 89:1010–28.

Sell, A., A. W. Lukazsweski, and M. Townsley. 2017. Cues of upper body strength account for most of the variance in men's bodily attractiveness. *Proceedings of the Royal Society B* 284:20171819.

Seyfarth, R. M., and D. L. Cheney. 2012. The evolutionary origins of friendship. *Annual Review of Psychology* 63:153–77.

Shaw, G. B. 1894. The religion of the pianoforte. *Fortnightly Review* 55 (326): 255–66.

Shell, J. 2019. *Giants of the Monsoon Forest: Living and Working with Elephants.* New York: Norton.

Silber, G. K. 1986. The relationship of social vocalizations to surface behavior and aggression in the Hawaiian humpback whale (*Megaptera novaeangliae*). *Canadian Journal of Zoology* 64:2075–80.

Silk, J. B. 1999. Why are infants so attractive to others? The form and function of infant handling in bonnet macaques. *Animal Behaviour* 57:1021–32.

Simmons, R. 2002. *Odd Girl Out: The Hidden Culture of Aggression in Girls.* New York: Harcourt.

Simpkin, T. 2020. Mixed feelings: How to deal with emotions at work. Total-jobs.com, January 8, 2020.

Slaby, R. G., and K. S. Frey. 1975. Development of gender constancy and selective attention to same-sex models. *Child Development* 46:849–56.

Slotow, R., et al. 2000. Older bull elephants control young males. *Nature* 408:425–26.

Small, M. F. 1989. Female choice in nonhuman primates. *Yearbook of Physical Anthropology* 32:103–27.

Smith, E. A., M. B. Mulder, and K. Hill. 2001. Controversies in the evolutionary social sciences: A guide for the perplexed. *Trends in Ecology and Evolution* 16:128–35.

Smith, T. M., et al. 2017. Cyclical nursing patterns in wild orangutans. *Science Advances* 3:e1601517.

Smith, T. W. 1991. Adult sexual behavior in 1989: Number of partners, frequency of intercourse and risk of AIDS. *Family Planning Perspectives* 23:102–7.

Smuts, B. B. 1985. *Sex and Friendship in Baboons.* New York: Aldine.

———. 1987. Gender, aggression, and influence. In *Primate Societies,* ed. B. Smuts et al., pp. 400–12. Chicago: University of Chicago Press.

———. 1992. Male aggression against women: An evolutionary perspective. *Human Nature* 3:1–44.

———. 2001. Encounters with animal minds. *Journal of Consciousness Studies* 8:293–309.

Smuts, B. B., and R. W. Smuts. 1993. Male aggression and sexual coercion of females in nonhuman primates and other mammals: Evidence and theoretical implications. *Advances in the Study of Behavior* 22:1–63.

Snowdon, C. T., and T. E. Ziegler. 2007. Growing up cooperatively: Family processes and infant care in marmosets and tamarins. *Journal of Developmental Processes* 2:40–66.

Sommers, C. H. 2012. You can give a boy a doll, but you can't make him play with it. *Atlantic,* December 6, 2012.

Spear, B. A. 2002. Adolescent growth and development. *Journal of the American Dietetic Association* 102:S23–29.

Spelman, E. V. 1982. Woman as body: Ancient and contemporary views. *Feminist Studies* 8:109–31.

Spinka, M., R. C. Newberry, and M. Bekoff. 2001. Mammalian play: Training for the unexpected. *Quarterly Review of Biology* 76:141–68.

Staes, N., et al. 2017. FOXP2 variation in great ape populations offers insight into the evolution of communication skills. *Scientific Reports* 7:16866.

Stanford, C. B. 1998. The social behavior of chimpanzees and bonobos. *Current Anthropology* 39:399–407.

Stavro, E. 1999. The use and abuse of Simone de Beauvoir: Re-evaluating the French poststructuralist critique. *European Journal of Women's Studies* 6:263–80.

Stern, B. R., and D. G. Smith. 1984. Sexual behaviour and paternity in three captive groups of rhesus monkeys. *Animal Behaviour* 32:23–32.

Stöckl, H., et al. 2013. The global prevalence of intimate partner homicide: A systematic review. *Lancet* 382:859–65.

Strum, S. C. 2012. Darwin's monkey: Why baboons can't become human. *Yearbook of Physical Anthropology* 55:3–23.

Stulp, G., A. P. Buunk, and T. V. Pollet. 2013. Women want taller men more than men want shorter women. *Personality and Individual Differences* 54:877–83.

Stumpf, R. M., and C. Boesch. 2010. Male aggression and sexual coercion in wild West African chimpanzees, *Pan troglodytes verus. Animal Behaviour* 79:333–42.

Stutchbury, B. J. M., et al. 1997. Correlates of extra-pair fertilization success in hooded warblers. *Behavioral Ecology and Sociobiology* 40:119–26.

Sugiyama, Y. 1967. Social organization of Hanuman langurs. In *Social Communication among Primates*, ed. S. A. Altmann, pp. 221–53. Chicago: University of Chicago Press.

Surbeck, M., and G. Hohmann. 2013. Intersexual dominance relationships and the influence of leverage on the outcome of conflicts in wild bonobos. *Behavioral Ecology and Sociobiology* 67:1767–80.

Surbeck, M., et al. 2017. Sex-specific association patterns in bonobos and chimpanzees reflect species differences in cooperation. *Royal Society Open Science* 4:161081.

———. 2019. Males with a mother living in their group have higher paternity success in bonobos but not chimpanzees. *Current Biology* 29:R341–57.

Swaab, D. F. 2010. *Wij Zijn Ons Brein*. Amsterdam: Contact (Dutch).

Swaab, D. F., and M. A. Hofman. 1990. An enlarged suprachiasmatic nucleus in homosexual men. *Brain Research* 537:141–48.

Taylor, S. 2002. *The Tending Instinct: How Nurturing Is Essential for Who We Are and How We Live.* New York: Henry Holt.

Thornhill, R., and C. T. Palmer. 2000. *A Natural History of Rape: Biological Bases of Sexual Coercion*. Cambridge, MA: MIT Press.

Tiger, L. 1969. *Men in Groups*. New York: Random House.

Titze, I. R., and D. W. Martin. 1998. Principles of voice production. *Journal of the Acoustical Society of America* 104:1148.

Tjaden, P., and N. Thoennes. 2000. *Full report of the prevalence, incidence, and consequences of violence against women*. U.S. Department of Justice, Office of Justice Programs.

Todd, B. K., and R. A. Banerjee. 2018. Lateralisation of infant holding by mothers: A longitudinal evaluation of variations over the first 12 weeks. *Laterality: Asymmetries of Brain, Body and Cognition* 21:12–33.

Todd, B. K., et al. 2018. Sex differences in children's toy preferences: A systematic review, meta-regression, and meta-analysis. *Infant and Child Development* 27:e2064.

Tokuyama, N., and T. Furuichi. 2017. Do friends help each other? Patterns of female coalition formation in wild bonobos at Wamba. *Animal Behaviour* 119:27–35.

Tokuyama, N., T. Sakamaki, and T. Furuichi. 2019. Inter-group aggressive interaction patterns indicate male mate defense and female cooperation across bonobo groups at Wamba, Democratic Republic of the Congo. *American Journal of Physical Anthropology* 170:535–50.

Townsend, S. W., T. Deschner, and K. Zuberbühler. 2008. Female chimpanzees use copulation calls flexibly to prevent social competition. *PLoS ONE* 3:e2431.

Tratz, E. P., and H. Heck. 1954. Der afrikanische Anthropoide "Bonobo," eine neue Menschenaffengattung. *Säugetierkundliche Mitteilungen* 2:97–101 (German).

Travis, C. B. 2003. *Evolution, Gender, and Rape*. Cambridge, MA: MIT Press.

Trivers, R. L. 1972. Parental investment and sexual selection. In *Sexual Selection and the Descent of Man*, ed. B. Campbell, pp. 136–79. Chicago: Aldine.

Troje, N. F. 2002. Decomposing biological motion: A framework for analysis and synthesis of human gait patterns. *Journal of Vision* 2:371–87.

Trost, S. G., et al. 2002. Age and gender differences in objectively measured physical activity in youth. *Medicine and Science in Sports and Exercise* 34:350–55.

Turner, P. J., and J. Gervai. 1995. A multidimensional study of gender typing in preschool children and their parents: Personality, attitudes, preferences, behavior, and cultural differences. *Developmental Psychology* 31:759–72.

Tutin, C. E. G. 1979. Mating patterns and reproductive strategies in a community of wild chimpanzees. *Behavioral Ecology and Sociobiology* 6:29–38.

Utami Atmoko, S. S. 2000. *Bimaturism in orang-utan males: Reproductive and ecological strategies.* Ph.D. thesis, University of Utrecht.

Vacharkulksemsuk, T., et al. 2016. Dominant, open nonverbal displays are attractive at zero-acquaintance. *Proceedings of the National Academy of Sciences USA* 113:4009–14.

Vaden-Kierman, N., et al. 1995. Household family structure and children's aggressive behavior: A longitudinal study of urban elementary school children. *Journal of Abnormal Child Psychology* 23:553–68.

van Hooff, J.A.R.A.M. 2019. *Gebiologeerd: Wat een Leven Lang Apen Kijken Mij Leerde over de Mensheid.* Amsterdam: Spectrum (Dutch).

van Leeuwen, E., K. A. Cronin, and D. Haun. 2014. A group-specific arbitrary tradition in chimpanzees (*Pan troglodytes*). *Animal Cognition* 17:1421–25.

van Schaik, C. 2004. *Among Orangutans: Red Apes and the Rise of Human Culture.* Cambridge, MA: Belknap Press.

van Woerkom, W., and M. E. Kret. 2015. Getting to the bottom of processing behinds. *Amsterdam Brain and Cognition Journal* 2:37–52.

Vasey, P. L. 1995. Homosexual behavior in primates: A review of evidence and theory. *International Journal of Primatology* 16:173–204.

Vauclair, J., and K. Bard. 1983. Development of manipulations with objects in ape and human infants. *Journal of Human Evolution* 12:631–45.

Verloigne, M., et al. 2012. Levels of physical activity and sedentary time among 10- to 12-year-old boys and girls across 5 European countries using accelerometers: An observational study within the ENERGY-project. *International Journal of Behavioral Nutrition and Physical Activity* 9:34.

Vines, G. 1999. Queer creatures. *New Scientist,* August 7, 1999.

Volk, A. A. 2009. Human breastfeeding is not automatic: Why that's so and what it means for human evolution. *Journal of Social, Evolutionary, and Cultural Psychology* 3:305–14.

von Rohr, C. R., et al. 2012. Impartial third-party interventions in captive chimpanzees: A reflection of community concern. *PLoS ONE* 7:e32494.

Voskuhl, R., and S. Klein. 2019. Sex is a biological variable—in the brain too. *Nature* 568:171.

Watts, D. P., F. Colmenares, and K. Arnold. 2000. Redirection, consolation, and male policing. In *Natural Conflict Resolution,* ed. F. Aureli, and F. B. M. de Waal, pp. 281–301. Berkeley: University of California Press.

Watts, D. P., et al. 2006. Lethal intergroup aggression by chimpanzees in Kibale National Park, Uganda. *American Journal of Primatology* 68:161–80.

Wayne, S. 2021. *Alpha Male Bible: Charisma, Psychology of Attraction, Charm.* Hemel Hempstead, UK: Perdens.

Weatherford, J. 2004. *Genghis Khan and the Making of the Modern World.* New York: Broadway Books.

Weidman, N. 2019. Cultural relativism and biological determinism: A problem in historical explanation. *Isis* 110:328–31.

Weisbard, C., and R. W. Goy. 1976. Effect of parturition and group composition on competitive drinking order in stumptail macaques. *Folia primatologica* 25:95–121.

Westneat, D. F., and R. K. Stewart. 2003. Extra-pair paternity in birds: Causes, correlates, and conflict. *Annual Review of Ecology, Evolution, and Systematics* 34:365–96.

Westover, T. 2018. *Educated: A Memoir.* New York: Random House.

White, E. 2002. *Fast Girls: Teenage Tribes and the Myth of the Slut.* New York: Scribner.

Wickler, W. 1969. Socio-sexual signals and their intra-specific imitation among primates. In *Primate Ethology,* ed. D. Morris, pp. 89–189. Garden City, NY: Anchor Books.

Wiederman, M. W. 1997. The truth must be in here somewhere: Examining the gender discrepancy in self-reported lifetime number of sex partners. *Journal of Sex Research* 34:375–86.

Williams, C. L., and K. E. Pleil. 2008. Toy story: Why do monkey and human males prefer trucks? *Hormones and Behavior* 54:355–58.

Wilson, E. A. 1998. *Neural Geographies: Feminism and the Microstructure of Cognition.* New York: Routledge.

———. 2000. Neurological preference: LeVay's study of sexual orientation. *SubStance* 29:23–38.

Wilson, E. O. 1978. *On Human Nature.* Cambridge, MA: Harvard University Press.

Wilson, M. L., et al. 2014. Lethal aggression in Pan is better explained by adaptive strategies than human impacts. *Nature* 513:414–17.

Wiseman, R. 2016. *Queen Bees and Wannabes: Helping Your Daughter Survive Cliques, Gossip, Boys, and the New Realities of Girl World.* New York: Harmony.

Wittig, R. M., and C. Boesch. 2005. How to repair relationships: Reconciliation in wild chimpanzees. *Ethology* 111:736–63.

Wolfe, L. 1979. Behavioral patterns of estrous females of the Arashiyama West troop of Japanese macaques. *Primates* 20:525–34.

Wrangham, R. W. 2019. *The Goodness Paradox: The Strange Relationship Between Virtue and Violence in Human Evolution.* New York: Pantheon.

Wrangham, R. W., and D. Peterson. 1996. *Demonic Males: Apes and the Evolution of Human Aggression.* Boston: Houghton Mifflin.

Yamamichi, M., J. Gojobori, and H. Innan. 2012. An autosomal analysis gives no genetic evidence for complex speciation of humans and chimpanzees. *Molecular Biology and Evolution* 29:145–56.

Yamamoto, S., T. Humle, and M. Tanaka. 2009. Chimpanzees help each other upon request. *PLoS One* 4:e7416.

Yancey, G., and M. O. Emerson. 2016. Does height matter? An examination of height preferences in romantic coupling. *Journal of Family Issues* 37:53–73.

Yerkes, R. M. 1925. *Almost Human.* New York: Century.

———. 1941. Conjugal contrasts among chimpanzees. *Journal of Abnormal and Social Psychology* 36:175–99.

Yong, E. 2019. Bonobo mothers are very concerned about their sons' sex lives. *Atlantic,* May 20, 2019.

Young, L., and B. Alexander. 2012. *The Chemistry Between Us: Love, Sex, and the Science of Attraction.* New York: Current.

Zahn-Waxler, C., et al. 1992. Development of concern for others. *Developmental Psychology* 28:126–36.

Zhou, J.-N., M. Hofman, L. Gooren, and D. F. Swaab. 1995. A sex difference in the human brain and its relation to transsexuality. *Nature* 378:68–70.

Zhou, W., et al. 2014. Chemosensory communication of gender through two human steroids in a sexually dimorphic manner. *Current Biology* 24:1091–95.

Zihlman, A. L., et al. 1978. Pygmy chimpanzee as a possible prototype for the common ancestor of humans, chimpanzees, and gorillas. *Nature* 275:744–46.

Zimmer, C. 2018. *She Has Her Mother's Laugh: The Powers, Perversions, and Potential of Heredity.* New York: Dutton.

Zuckerman, S. 1932. *The Social Life of Monkeys and Apes.* London: Routledge and Kegan Paul.

———. 1991. Apes are not us. *New York Review of Books,* May 30, 1991, pp. 43–49.

INDEX

Pages in *italics* refer to illustrations, photographs, and tables.

ABOUT THE AUTHOR

Frans de Waal is a Dutch-American ethologist and primatologist born in 1948 in Den Bosch, the Netherlands. He studied at the Universities of Nijmegen and Groningen before receiving a Ph.D. in biology from Utrecht University, in 1977. His first book, *Chimpanzee Politics* (1982), compared the schmoozing and scheming of power-hungry apes with that of human politicians.

In 1981 he moved to the United States to work at the Wisconsin National Primate Research Center in Madison. In 1992 he joined the psychology department of Emory University, in Atlanta, where he became C. H. Candler Professor. He was director of the Living Links Center at the Yerkes National Primate Research Center. Later he was also appointed Distinguished Professor at Utrecht University. Since 2019 he has been emeritus professor at both universities.

With his discovery of reconciliation behavior among primates, de Waal pioneered research on animal conflict resolution. In 1989 he received the Los Angeles Times Book Prize for *Peacemaking Among Primates*. During the same period, he studied bonobos and was the first to introduce this erotic species to the general public. Since then de Waal's name has become associated with animal empathy, emotions, and cooperation.

De Waal's research team has published hundreds of scientific articles in journals ranging from *Science*, *Nature*, and *Scientific American* to those specializing in animal behavior and cognition. His popular books—translated into over twenty languages—have made him one of the world's most visible biologists. His latest two are: *Are We Smart Enough to Know How Smart Animals Are?* (2016) and *Mama's Last Hug: Animal Emotions and What They Tell Us about Ourselves* (2019).

De Waal has been elected to the U.S. National Academy of Sciences as well as the Royal Netherlands Academy of Arts and Sciences. In 2007 *Time* voted him one of the World's 100 Most Influential People Today. In 2020 *Mama's Last Hug* received the PEN/E.O. Wilson Literary Science Writing Award.

He lives with his wife, Catherine Marin, in a forested area filled with wildlife in Smoke Rise, Georgia.